Plants, Gene and Crop Biotechnology

Plants, Gene and Crop Biotechnology

Brindra Chauhan

Plants, Gene and Crop Biotechnology

ISBN 978-93-5111-601-1

Published in 2015 in India by

RANDOM PUBLICATIONS

4376-A/4B, Gali Murari Lal, Ansari Road
New Delhi-110 002
Phone : +9111-43580356, 011-23289044, 011-43142548
e-mail: sales@randompublications.com,
info@randompublications.com, randomexports@gmail.com

Type Setting by : Friends Media, Delhi-110089
Digitally Printed at: Replika Press Pvt. Ltd.

Preface

Just five decades have passed since Watson and Crick deciphered the double helix structure of DNA, the stuff of genes, and laid the groundwork for modern biotechnology. Today we know the full genetic sequence of more than sixty living organisms. The list includes human beings, over thirty important human pathogens, the photosynthetic plant Arabidopsis thaliana, plant pathogens, several Archaea from exotic environments such as ocean depths and hot springs, and nature's own genetic engineer, the bacterium Agrobacterium tumefaciens, which for millennia has been transferring bacterial genes into plants to create new plant traits Agrobacterium finds useful. In 2001, more than 5.5 million farmers worldwide planted about 52.6 million hectares of crops that were genetically manipulated. This year the area has expanded further with India and Indonesia joining twelve other countries, including China, Mexico, South Africa, and Argentina, in approving the commercial planting of GM crops. But, agricultural biotechnology means a lot more than the creation of GM crops. It also involves the use of tissue culture to rapidly propagate disease- free seedling plants and to create new hybrids between plants that do not cross naturally, the use of sophisticated DNA-based genetic markers that allow breeders to follow and select for important traits more easily, and the use of DNA chips and other DNA-based diagnostic techniques to characterize pathogen populations for more effective deployment of resistant varieties.

I would like to thank my team for standing beside me throughout my career and writing this book. My special thanks go to "Random Publications" who have published the book.

– Brindra Chauhan

Contents

1

Plant Structure

INTRODUCTION

A plant has two organ systems: 1) the shoot system, and 2) the root system. The shoot system is above ground and includes the organs such as leaves, buds, stems, flowers (if the plant has any), and fruits (if the plant has any). The root system includes those parts of the plant below ground, such as the roots, tubers, and rhizomes.

The points on the stem to which the leaves are attached are usually slightly thickened, and are called nodes or 'joints' the lengths of stem between them being termed internodes. It has been already noticed that the seedling bean plant consists of a descending portion, the root, and an ascending part which comes above ground.

The latter is known as the primary shoot, and consists of an axis—the stem—upon which are arranged a series of lateral appendages which are called leaves. Flowers ultimately arise upon the shoot, and it is one of the characteristics of seed-bearing plants that seeds are always produced on their shoots and never on roots.

For the present, however, flowers may be left for future consideration, and attention paid to the origin and nature of the vegetative shoot or stem with its ordinary green leaves. In the earliest stages of the development of a bean plant the primary shoot is very short and bears the cotyledons or primary leaves, its tip ending in the plumule.

The latter is a bud, and at the time when the seed commences to germinate, its parts cannot be fully made out by observations with the naked eye. As soon as it comes above ground, however, the bud is found, on examination, to consist of a short stem hidden by a number of enfolding leaves. As growth proceeds, the short stem inside the bud elongates, and the leaves, which at first are crowded upon it, become separated from each other.

Marks made upon the stem as previously explained for the root, reveal the fact that the increase in length takes place at the tip of the shoot. After reaching a certain length the lowest intervals between the leaves cease to elongate; the upper, younger and shorter ones also lengthen and cease in a similar manner, to be followed in turn by still younger parts of the stem nearer the tip. The stem, before the growing season is over, may thus reach a height of two or three feet, or even more, the extreme tip, or growing-point as it is called, remaining young all the time, and acting as a manufactory for the addition of more stem and leaves.

The growing-point, which is of a tender and delicate nature, is protected by the enfolding young leaves, the latter arising as outgrowths from its external surface.

The youngest leaves are always nearest the tip the stem which bears them, the older ones being further removed from it in regular order—that is, they arise in acropetal succession, and adventitious leaves are never met with. Instead of the latter growing at once into a long shoot, bearing leaves at some distance from each other, the primary axis within the plumule elongates very little, its internodes remain very short, and the leaves arising upon it appear crowded together, usually in the form of a rosette, a short distance above where the cotyledons were placed; this form of stem with short contracted inter-nodes is well illustrated in the first season's growth of mangels, turnips, carrots, certain thistles, and red clover.

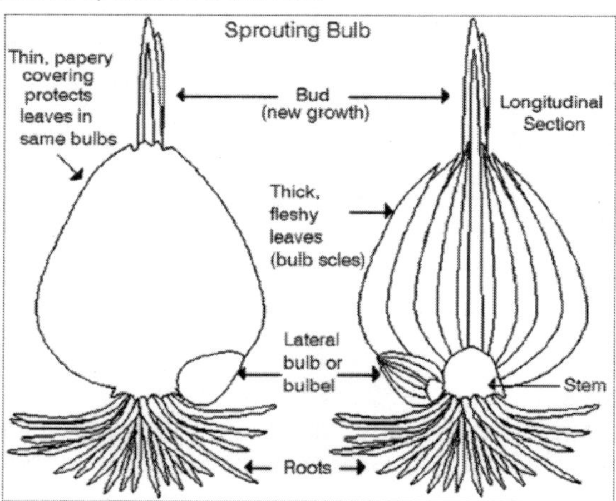

Fig. Growing Stem of Onion

In such plants as these, the primary root and hypocotyl become much thickened by the deposition within them of reserve-food prepared by the leaves, and it is only during the following year that the growing-point of the stem, which is hidden in the centre of the rosette, elongates and produces a shoot with long internodes, and bearing a series of new leaves at considerable intervals.

In the onion and many bulbous plants the primary stem also remains very short, and the reserve-foods prepared by them are deposited in the bases of the leaves, instead of being stored in the root and stem, as in the former instances.

THE ESSENTIAL PARTS OF THE FLOWER

The andrcecium and gynaecium are directly concerned in the production of seed, as explained hereafter, and are termed the essential parts of a flower. The Andrcecium consists of stamens, each of which, is a modified form of leaf, although its appearance and structure is very different from the petals and sepals of the perianth.

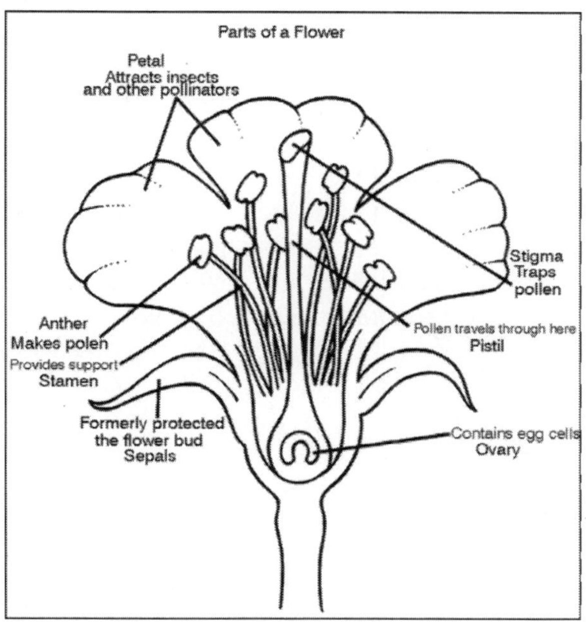

Fig. Parts of Flower

A stamen usually consists of a more or less elongated threadlike portion — the filament — surmounted by a swollen thicker part termed the anther. The anther consists of two somewhat elongated halves or anther-lobes, which are situated usually on opposite sides of the upper part of the filament: the part of the filament uniting the anther-lobes is termed the connective. Most frequently the stamens of the androecium are distinct and completely free from each other as in the buttercup, but in some flowers the filaments of the stamens are united together and only the anthers are free.

When all the filaments are united the stamens are described as monadelphous; if two or several separate bundles of united filaments are present the stamens are said to be diadelphous and polyadelphous respectively. In the daisy, dandelion, and most plants belonging to the Composite, the anthers are united and the filaments are free; such stamens are termed syngenesious. Stamens attached to the petals, as in the potato flower, are described as epipetalous. The gynaecium is composed of carpels, each of which generally consists of three parts:

- A swollen hollow basal portion termed the ovary,
- A thin more or less elongated part called the style,
- The stigma.

Within the cavity of the ovary are small round or oval bodies termed ovules, which under certain circumstances develop into seeds. The part inside the ovary on which the ovules are borne is termed the placenta. The carpel may be considered as a leaf which has been folded along the midrib and united at its edges.

The line corresponding to the united edges of the leaf is termed the ventral suture of the carpel, and it is along this line that the ovules are generally attached in two rows—one row belonging to each edge; the line corresponding to the mWrib of the folded leaf is the dorsal suture.

These parts are readily seen in the pod of a pea, which bears considerable resemblance to a folded green leaf. The gynaecium may consist of separate carpels as in the buttercup, in which case it is said to be apocarpous. Frequently the carpels are united and then form what is termed a syncarpous gynaecium. The amount of union among the carpels varies, but very frequently their ovaries are completely united to form one common ovary: in such cases the styles are generally united to form one common style, the corresponding stigmas usually remaining free. When the carpels of the syncarpous gynaecium are united by their edges as at the ovary possesses only one cavity or loculus and is said to be unilocular. In other examples the carpels are folded so that their edges meet in the middle of the ovary, the united parts forming partitions or dissepiments dividing up the common ovary into several cavities such ovaries are described as multilocular and each loculus corresponds to a single carpel.

Occasionally the number of loculi insftie an ovary does not correspond with the number of carpels present in the latter, as dissepiments occur which are not formed from the united walls of two neighbouring carpels but which are produced by the growth inwards of a portion of the ovary wall. The latter are termed false dissepiments, an example of which is the septum which divides the ovary in the Cruciferae.

NON-ESSENTIAL PARTS OF THE FLOWER

The calyx and corolla whorls of floral leaves together constitute the perianth of the flower, and as they are not directly concerned in the production of seeds are termed the non-essentialparts of the flower. When one of the whorls of the perianth is absent as in the mangel, male hop, and anemone, the flower is spoken of as monochlamydeous; if both calyx and corolla are absent, as in the ash and willow, the flower is naked or achlamydeous.

THE CALYX

The calyx forms a protective covering for the rest of the flower when the latter is still young, and may either lall off when the flower opens, in which case it is caducous ^ or remain attached to the receptacle for an indefinite period, when it is described as a persistent calyx. It is usually green but may assume some other colour, in which case it is spoken of as petaloid. A calyx which consists of free separate sepals, as in the buttercup, is termed polysepalous; those in which the sepals are united, as in the primrose and pea, are said to be gamosepalous. In groundsel, thistle, and other plants belonging to the Compositae, the calyx takes the form of a ring of hair known as a pappus, which generally develops rapidly after the corolla has faded and acts as a float for the distribution of the seed-case by means of the wind.

The Corolla

This part of the flower is usually of bright colour and serves mainly as an attraction for insects. When the petals forming it are free from each other, as in the buttercup and rose, the corolla is polypetalous; the term gamopetalous is applied to corollas which are composed of united petals, as in the primrose and Canterbury bell.

PLACENTATION

The arrangement of the placentas or points from which the ovules arise inside an ovary is termed placentation. When the ovules are arranged in lines on the wall of the ovary, the placentation is parietal.

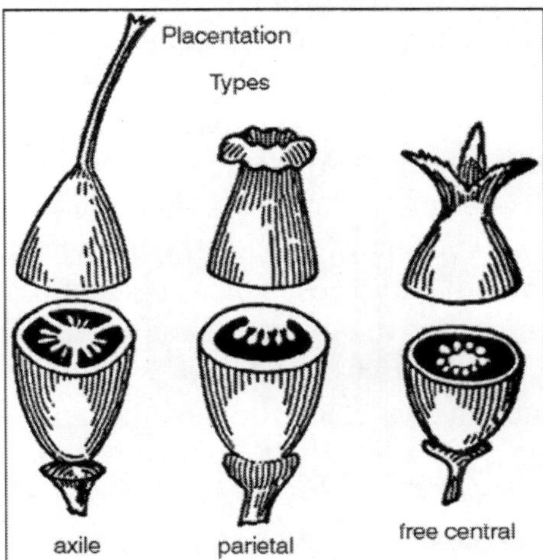

Fig. Placentation

In multilocular ovaries, such as at the ovules are generally arranged in the angles formed at the centre where the edges of the carpels are united, and the placentation is described as axile. In the primrose and chickweed families of plants the ovules are attached to a placenta which arises in the form of a short column from the base of the ovary and has no connection with the sides: this arrangement is known as free central placentation.

MONOCLINOUS AND DICLINOUS FLOWERS

When both the essential parts are present in the same flower, as in the buttercup, charlock, and the majority of common plants, the flower is described as monoclinous; sometimes the terms perfect, hermaphrodite or bisexual are applied to such flowers. In certain flowers, as those of the cucumber, melon, hop, hazel, and willow, one or other of the essential parts are missing: such are said to be diclinous imperfect or unisexual. Diclinous flowers may be of two kinds:

- Those in which the andrcecium is alone present and described as staminate or male flowers,
- Those in which only the gynaecium is met with and spoken of as carpellary, pistillate or female flowers.

When both kinds of diclinous flowers are met with on the same individual plant, as in the case of the cucumber and hazel, the plant is said to be moncecious; in examples, such as the hop and willow where the two kinds of diclinous flowers are produced on separate individuals, the plants are spoken of as dioecious.

BUDS

In botany, a bud is an undeveloped or embryonic shoot and normally occurs in the axil of a leaf or at the tip of the stem. Once formed, a bud may remain for some time in a dormant condition, or it may form a shoot immediately. Buds may be specialized to develop flowers or short shoots, or may have the potential for general shoot development. The term bud is also used in zoology, where it refers to an outgrowth from the body which can develop into a new individual.

The stems and leaves of all flowering plants originate from buds in the manner indicated above; buds may therefore be termed embryonic or incipient shoots. It is by their growth that trees, which appear so bare in winter, become clothed with fresh green leaves in the succeeding spring. The relationship which they bear to the leaves and stems produced by them is easily discerned by examining the structure and watching the development of the terminal bud of a young sycamore tree.

On the outside is observed a series of scaly leaves, which overlap each other, and protect and cover up the delicate growing point of the twig. The disposition of these scaly leaves and within are also seen the ordinary leaves arranged upon a very short stem (s). In spring the inner scaly leaves grow for a time.

The stem which bears the rudimentary green foliage-leaves, elongates, and the latter are pushed out from between the protective scaly leaves of the bud. After a week or ten days, the stem has reached a considerable length, and the leaves, which were rudimentary and packed away in the bud, unfold themselves and grow out flat. The number of foliage-leaves present upon a developed shoot is often indicated in the bud, but in some plants, especially those of a herbaceous character, the growing point of the bud continues to produce new leaves until frost checks it in autumn. The vegetative shoots of plants usually end in terminal buds and an examination of almost any kind of plant shows that not only are buds present at the tips of the stems, but on their sides as well. These lateral buds arise ordinarily in the upper angles formed where the leaf-bases and stem join each other. The angles are termed the axils of the leases, and the buds are designated axillary buds.

Most frequently only one bud is produced in each leaf-axil, but in some instances, two or more may be present. Generally the first leaves of the bud which are outermost or lowest down on the stem, are rudimentary structures, smaller and different in appearance from those which unfold later.

In the primary bud or plumule of the bean, and many similar herbaceous plants, this is observable, but it is most evident in buds which are met with upon perennials, such as shrubs and trees. In the latter the outermost leaves of the buds are generally more or less firm, tough structures, termed scales or scale-leaves, which protect the interior of the bud from being injured by frost, rain, and other agents during the winter.

Buds, such as those of the sycamore and pear, having scales are termed scaly buds, those without, such as mealy guelder-rose, being known as naked buds. Buds similar to those of the bean and sycamore, previously described, which develop into shoots bearing green foliage-leaves, are termed leaf-buds: when met with upon trees they are sometimes named wood-buds, as it is from them that new woody twigs are produced.

Many buds, however, on opening, give rise to flowers only, and are termed flower-buds: a third kind is met with producing short shoots bearing both green leaves and flowers; these are mixed-buds.

Among gardeners the two latter forms are known as fruit-buds as it is from them that fruit is obtained. It is not possible in all cases to distinguish fruit-buds and wood-buds by their outward appearance, although for budding and pruning operations and the general management of fruit-trees it is desirable to do so. In apples and pears the wood-buds are small and pointed, the fruit-buds being blunter, more plump, and of larger size.

In cherries and plums both kinds are very similar in appearance, and it is only in spring when they begin to develop that the stouter and blunter characters of the fruit-buds show themselves. Their position upon the shoots is a great aid in distinguishing the two classes of buds.

ADVENTITIOUS BUDS

Dormant buds, are buds which have arisen in regular order in the axils of leaves, but which have remained inactive some time; the only irregularity about them is their period of development. Buds may, however, arise at any point of a plant, not necessarily in the axil of a leaf, but on any part of the stem, or even upon roots and leaves: such are termed adventitious buds. Examples are met with on the roots of docks, poplars, roses, and many other plants, especially when the upper bud-bearing parts have been removed.

They frequently arise and produce shoots upon stems which have been injured. In some instances they proceed from the callus covering the wounds where branches have been cut off; some of the shoots of ' pollard ' trees spring from adventitious buds originating in this manner.

Adventitious buds are often produced upon leaves which have been removed from the parent and pegged down on moist sand or loam.

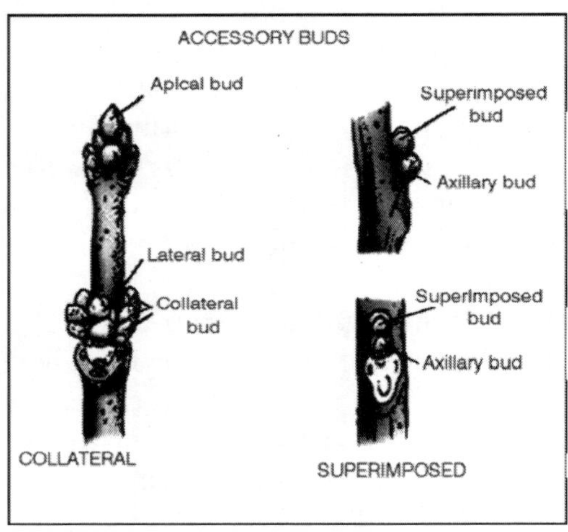

Fig. Adventitious Buds

Gardeners take advantage of this peculiarity in propagating begonias. Similar buds occur upon some kinds of leaves when they are severed from the plant and their petioles stuck in moist ground: the scales of the hyacinth and other bulbs give rise to new plants in this manner.

DORMANT BUDS

On examining trees in spring when the buds are beginning to develop, it will be observed that some of them remain inactive and continue in this condition all the summer.

Not only may they refuse to grow in what may be termed their proper season, but they frequently remain undeveloped for long periods. Such buds are termed dormant or resting buds, and are met with upon almost all kinds of plants, chiefly near the base of the stems.

Although many dormant buds soon die, some remain capable of development for several years after their formation, and may give rise to what are termed deferred shoots.

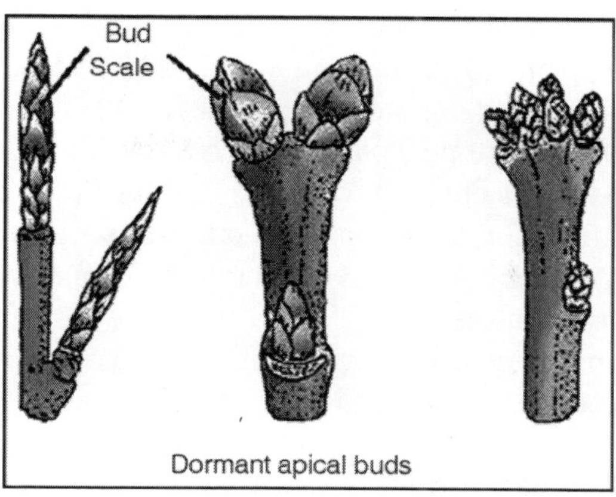

Fig. Dormant Buds

In fruit-trees they are termed c water-sprouts' j if they spring from beneath the surface of the ground they are known as ' suckers.' They not uncommonly arise upon 'stocks' which have been grafted or budded. Destruction of the terminal and lateral buds near the top of a stem tends to promote the growth of deferred shoots from dormant buds at its base.

Moreover, pinching out the terminal buds of herbaceous and other plants is often practised with a view to insure the development of all the lateral buds upon it, and the formation of a bushy plant instead of one with a single main stem and few branches.

The grazing and mowing of grasses promotes the full development of all their buds, and a consequent increase of leafy stems. Not only does cutting away or pinching flff the terminal buds promote the development of basal buds likely to become dormant, but anything which impedes the movement of water or 'flow of sap' to the terminal and highly-placed buds tends towards the same result.

In the early formation of cordon fruit-trees, where it is important that all the buds upon the main stem should develop shoots or short spurs, bending the shoot for a time is practised in order to promote the ' breaking' of those buds at the base of the stem which might otherwise remain dormant and leave a length of unfruitful wood.

STEMS AND THEIR VARIETIES

Stem usually consist of three tissues, dermal tissue, ground tissue and vascular tissue. The dermal tissue covers the outer surface of the stem and usually functions to waterproof, protect and control gas exchange. The ground tissue usually consists mainly of parenchyma cells and fills in around the vascular tissue. It sometimes functions in photosynthesis. Vascular tissue provides long distance transport and structural support. Most or all ground tissue may be lost in woody stems. The dermal tissue of aquatic plants stems may lack the waterproofing found in aerial stems. The arrangement of the vascular tissues varies widely among plant species.Stems which are soft and which usually last but a short time, are termed herbaceous practically all our annuals have stems of this nature, and many perennial plants also, *e.g.* nettles and hops. Most stems which last several seasons develop within themselves considerable quantities of wood, and are harder and firmer in consequence: such stems are said to be woody. It must be pointed out, however, that herbaceous stems in reality also possess wood, but only in the form of thin strands, which are relatively small in amount when compared with the remaining soft parts.

All stems, moreover, are soft and herbaceous when very young, so that no real distinction exists between herbaceous and woody stems, as it is a matter of degree of development of the wood within them: a wall-flower or a rose, may be soft and herbaceous in its upper parts and hard and woody below.

Trees and Shrubs have well-developed woody stems, the former possessing a single main stem or trunk, which is devoid of branches for some distance above the ground; the latter have no very distinct main stem, and the chief branches are all much the same in thickness and spring from a point either on or close to the ground.

Many plants have stems which are too weak to maintain an erect position; they consequently grow along the surface of the soil. Some plants have weak stems which always remain prostrate while others, designated climbing plants ^ have stems which, although too weak to stand upright of themselves, are nevertheless able to use suitable objects near them as supports. Climbing plants support themselves in various ways. In ivy, adventitious roots are developed on one side of the stem, and these serve to fix the plant to bark of trees, walls, and rocks.

Tropaeolums of gardens and wild clematis are supported by their leaves, the petioles of which curve round the stronger branches of plants growing near them. Peas and vetches are also enabled to climb by means of their leaves, some of the leaflets of which are modified into thin threadlike structures termed tendrils. The latter are sensitive to contact, and wind round any slender object which they touch. Plants such as the blackberry, rose, are supported by means of their stiff prickles. In twining plants the whole stem upholds itself by twisting round neighbouring objects.

The stems of some of them always twine to the right when growing round a support; the hop is an example: others, such as bindweed, twine to the left. A number of peculiar modifications of shoots are met with,many of which receive special names: In the wild pear, wild plum, hawthorn, sloe, and buckthorn, some of the branches end in hard, sharp points, termed thorns or spines. That they are modified shoots is seen from the fact that they arise in the axils of leaves, and also themselves bear leaves and lateral buds in some instances. A runner or stolon is a shoot which extends horizontally over the surface of the ground.

Its internodes are long, and from its nodes adventitious roots are produced, and grow into the soil. The buds present on the runner then become fixed to the ground, and, developing into upright shoots, form separate plants as soon as the internodes at s die away or are severed.

Fig. Strawberry Runners

Strawberry runners and those of creeping crowfoot are good examples. Stems within the soil sometimes resemble roots, but they can be distinguished from the latter by the possession of leaves and buds, and by their originating in the axils of leaves. A rhizome or ' rootstock ' is an underground shoot, which grows more or less horizontally. Adventitious roots arise at the nodes, and the internodes may be long or short, thick or thin.

Most rhizomes are, however, definite in growth, the main axis, after growing a longer or shorter distance below, comes above ground, the continuation of the rhizome within the soil being carried on by lateral branches. In perennial rhizomes of definite growth, such as those of sedges, grasses, and many other plants, the permanent part which remains below ground is a false main-axis or sym-podium. The term sucker is applied to any adventitious shoot which originates below ground on the stems or roots of shrubs and trees. It possesses adventitious roots and by separation from the parent may become a new individual plant.

Suckers often develop very rapidly and rob the parent of water and nutriment, so that except for purposes of propagation they he onion seedling, develops several leaves during summer and the plant swells at its base and forms a bulb. Tracing the leaves from the green parts downward, it is observed that the bases, especially of the inner ones, are thickened, and it is these leaf-bases which form the main mass of the bulb, the stem upon which they grow being comparatively small. At number of thick, fleshy, scale-leaves, which overlap each other more or less completely.

The whole structure is practically a huge bud, and in the axils of some of its scales are small, rudimentary buds. Familiar examples are met with in the onion, tulip, lily, hyacinth, snowdrop, and narcissus. The onion seedling develops several leaves during summer and the plant swells at its base and forms a bulb. Tracing the leaves from the green parts downward, it is observed that the bases, especially of the inner ones, are thickened, and it is these leaf-bases which form the main mass of the bulb, the stem (s) upon which they grow being comparatively small.

At the end of summer, the green parts of the leaves die and shrivel; their lower parts, which have become thin, act as a cover for the rest of the bulb, and prevent the rapid loss of water from the interior.

The onion bulb, if planted next year, forms adventitious roots from the base of the stem, and the terminal growing-point inside grows up into the air, and produces leaves and an inflorescence of white flowers at the end of a long hollow stem. The buds in the axils of the scale leaves develop usually in the same manner, so that from one bulb several flowering shoots are often produced.

The materials stored in the bulb-scales are used up in this development of the flowering stems, and after the production of ripe seeds, the whole plant is generally exhausted, and dies away, in which case the onion is a biennial plant.

Occasionally, however, some of the lateral buds from the axils of the scales do not produce inflorescences, but leafy shoots only, which form small bulbs in the same manner as an onion seedling. After the death of the parent, these smaller bulbs remain, and carry on the growth in the succeeding year. The onion plant in this instance is a perennial. A tulip bulb in autumn consists of a short, thick stem, upon which are placed a series of large, overlapping, fleshy scales.

The latter are complete leaves, and not merely leaf bases, as in the onion. At the apex of the stem is an embryonic shoot, having leaves upon it, and bearing a terminal flower; in the axils of some of the scales are rudimentary buds. In spring the flower-bearing stem grows from within the bulb and comes above ground, carrying with it the flower and two or three leaves. This development takes place at the expense of the food stored in the scales: the latter therefore soon become soft, and at the end of the season shrivel up and decay. The leaves on emerging from the soil turn green, and during the spring and summer manufacture a consfderable amount of food; that part of it not needed for the plant's immediate requirements is transferred to lateral axillary buds below ground, and is there stored.

These buds consequently grow rapidly and become young daughter bulbs; one of them is shown at n in course of development. Bulbs like the onion, tulip, and hyacinth, which have broad, concave scales arranged in such a manner that the outer ones completely enclose those within, are known as tunicated bulbs. In lilies the bulb scales are not so broad, and are arranged to overlap each other like the tiles on a roof: such bulbs are said to be imbricated.

THE LEAF

Typically a leaf is a thin, flattened organ borne above ground and specialized for photosynthesis, but many types of leaves are adapted in ways almost unrecognisable in those terms: some are not flat (for example many succulent leaves and conifers), some are not above ground (such as bulb scales), and some are without major photosynthetic function (consider for example cataphylls, spines, and cotyledons).

Although to a casual observer the leaves appear to be without any regular arrangement upon a plant, careful inspection shows that they are distributed on the stems in a very definite order, which is usually constant for each species. In some, such as the sycamore, dead-nettle, and cleavers, two or more leaves arise at the same node of the stem.

Each collection of leaves is then called a whorl and the individuals comprising it are always separated from each other by regular angular intervals. Thus, if two leaves are present they are half the circumference of the stem apart or exactly opposite each other, and not both on the same side; if three arise at the same node, they are separated from each other by regular intervals of 120 degrees, or one-third of the circumference, and so on for any number of leaves. On many stems the leaves are not in whorls but scattered singly along it, only one leaf arising at each node: such an arrangement is spoken of as alternate or spiral.

A line drawn from the bottom to the top of a shoot in such a manner that it touches the base of each successive leaf is a spiral. The distances between the leaves measured along the stem are variable, some being an inch apart, others two or more; their angular intervals apart are, however, as definite and regular as in plants with the whorled arrangement. The divergence or angular distance is usually expressed in fractions of the circumference. In elm, Spanish chestnut and grasses, it is that is, the spiral in passing from one leaf to the next winds half round the stem. In birch it is “J” while in pear and plum the angular distance is of the circumference. If any particular leaf in a row is selected and the spiral traced round the stem touching each successive leaf until another leaf is reached on the same row, the number of leaves touched, not counting the one at which we start, is equal to the number of the denominator of the fractions expressing the angular divergence, and the numerator indicates the number of complete turns round the stem which the spiral line traces.

For example, the angular divergence of the leaves on a pear shoot is selecting any one leaf as a starting point, the spiral line passes twice round the stem by the time that it reaches the next leaf on the same row, and in doing so touches the bases of five leaves. To determine the leaf-arrangement upon any particular shoot, it is necessary to observe the bases of the leaves and not the blades, as the position of the latter is affected by external conditions, especially by light and the force of gravitation. Occasionally the stems become twisted during growth, and the leaves are consequently displaced from their normal position. The orderly arrangement of the leaves upon stems is de pendent on the internal forces of the living plant. By growing in this manner all the leaves become equally exposed to light and air, and interfere very much less with each others requirements in this respect, than would be the case if the leaves were disposed irregularly.

FOLIAGE-LEAF

Those which are most conspicuous upon plants are green and are designated foliage-leaves. They are important organs generally concerned with the manufacture of food needed by the growing part of the plant, and are also organs from which much of the water taken from the soil by the roots is given off into the air.

A typical green foliage-leaf consists of the following parts—

- A broad expanded portion termed the blade or lamina.
- A slender stalk or petiole;
- A somewhat flattened basal sheath which connects the leaf to the stem.

The leaf-sheath often bears two appendages—the stipules—which may be broad and wing-like as in the clovers and pea, or small and narrow as in pear and apple; leaves possessing them are said to be stipulate, while those without are exstipufate. The parts of the leaf are of very varied form. In the grasses the sheath completely embraces the stem, and in the Umbelliferae it is very prominent; in many plants it is scarcely visible.

The petiole where present is usually narrow and cylindrical; frequently it is very short or missing altogether, in which case the leaf is described as sessile. The blade is generally the most obvious part of a foliage-leaf and the points of importance to notice at present are its venation, outline, margin, apex, and character of its surface.

Venation of Leaf-blade

The substance of the leaf is traversed by a number of woody strands which are termed veins or nerves, although it must not be inferred that they are similar in structure or function to the veins or nerves of animals. The arrangement of these strands is termed the venation of the leaf, of which there are two common types, namely

- Parallel
- Reticulate or net-venation.

In the first type the chief strands all run parallel to each other from the base of the leaf to the tip, as in the leaves of grasses, onion, hyacinth, lily-of-the-valley, and Monoco-tyledons generally. In net-veined leaves the smaller very delicate strands form a fine net-work within the leaf and this arrangement is characteristic of Dicotyledons. Of reticulate veined leaves two divisions are made according to the arrangement of the main strands. In one, the leaves have a central strand or mid-rib running down the middle of the leaf and from it are given off slightly smaller branch strands; such leaves are pinnately veined or feather-veined, those of the apple, plum, and peach are good examples. In the other division each leaf has several strong strands which start from the base of the blade and spread across to its margins somewhat like the fingers of an outstretched hand; such a leaf is described palmatdy veined. Ivy, sycamore, and currant leaves show this type of venation.

Forms of Blade

The outline of the blade of the leaf may assume almost any geometrical figure. When it is much elongated and narrow as in grasses it is termed a linear leaf.

It may also be lanceolate as in the narrow-leaved plantain; ovate egg-shaped); elliptical; reniform or kidney-shaped; cordate (heart-shaped); sagittate (arrow-shaped); spathulate (spoon-shaped) as in the daisy; and hastate (halberd-shaped) as in sheep's sorrel.

PLANT GENOME STRUCTURE

The genomes of all eukaryotic species consist of single-copy, middle repetitive and high copy number sequences. To gain an understanding of the per centages of each of the classes within any particular species, a group of random clones can be hybridized to blots of plant DNA. One such study was performed by Zamir and Tanksley. They hybridized 50 random genomic clones to tomato DNA. Washing was performed at two stringencies. The following are the results with regards to copy number.

Clone Class	*Low Stringency Wash*	*High Stringency Wash*
Single copy clones	44%	78%
Multiple copy clones	46%	18%
Repetitive clones	10%	4%

The table shows that hybridization stringency has a significant effect upon the number of sequences to which a sequence hybridizes. At the higher stringencies, most of the clone recognized only a a single copy within the genome. What this also shows is that the tomato genome contains many sequences that are about 80% homologous, but fewer sequences that are highly homologous. Clearly similar sequences must have diverged by some mechanism during the evolution of the species. The authors also hybridized these 50 clones to filters containing tomato, related tomato species, eight Solanum species and one member of the Curcurbitaceae family.

As a control,single copy tomato cDNA clones were also hybridized to these clones. At moderate stringency, the cDNAs hybridized to all the tomato species, and most hybridized to the Solanum species. 80% of the clones hybridized to the related Curcurbitaceae species. In contrast, only 50% of the random clones hybridized to related tomato and Solanum species, and only 10 % hybridized to the Curcurbitaceae species. The principal conclusion that can be drawn from these hybridizations is that the random sequences (as represented by the random genomic clones) are evolving faster than the single-copy sequences. Why? At the higher stringency, homology to distant related species was reduced. Therefore these sequences have undergone greater divergence. This evidenced by the fact that 0/5 of the repetitive sequences hybridized to tobacco, whereas 20/45 of the single and multiple copy clones hybridized to tobacco.

EVOLUTIONARY RELATEDNESS

The typical experiment begins with a large collection of cultivars or species. These samples are then analysed by hybridization with RFLP clones or more recently by the PCR-RAPD amplification. The similarity of each pair of samples is measured by calculating the number of common bands or amplification products. One estimator was developed by Nei and L. The formula is:

$$F = 2nXY/(nX + nY)$$

where nX and nY are the total number of fragments for sample X and Y and nXY is the number of fragments shared by the samples. Data is normally collected for a relatively large number of hybridizations or PCR-RAPD amplifications. The similarity data is then used in a cluster analysis to develop dendrograms which show the molecular relatedness of the species.

These types of analyses can:

- Provide independent support to previous phylogenetic and evolution hypotheses.
- Identify gene pools within a genus.
- Estimate genetic diversity within a genus.
- Provide necessary data to select appropriate parents for a molecular mapping project.

PLANT GENOME ORGANIZATION AND STRUCTURE

Eukaryotic genomes are much more complex than prokaryotic genomes. And further, plant genomes are more complex than other eukaryotic genomes. Prior to the development of recombinant DNA technology genomes, were analysed by reassociation kinetics techniques. Reassociation kinetic experiments are performed by melting DNA and allowing it to reanneal upon itself or with another population of either DNA or RNA molecules. The kinetics of the reassociation provide data that can be used to analyse the overall structure, evolution and expression of genomes. The most common method of denaturing duplex DNA is by heating to 100°C. But how can we monitor this denaturation and subsequent renaturation? The most common method is by measuring the absorbance change in the ultraviolet region at 260 nm. The important property is that melted, single-stranded DNA absorbs about 40% more at 260 nm than duplex DNA. If we slowly heat DNA the absorbance will increase dramatically over a short range of temperatures. The mid-point of this transition is called the melting temperature or Tm. Under physiological conditions, the Tm usually lies in the range of 85-95°C. Thus, without altering the cellular conditions, the duplex DNA is stable in the cell. The exact temperature that a particular DNA melts depends on several parameters. The GC content is important because GC base pairs have three hydrogen bonds compared to the two for AT base pairs. Thus, the higher the GC content the higher the Tm.

Table. Effect of GC Content on Tm

%GC	*Melting Temperature*
40	87°C
60	95°C

DNA denaturation is reversible, and the reversible process is called renaturation. This process requires that the temperature be lowered gradually.

As this lowering of temperature proceeds the following occurs:

- Single-stranded molecules randomly encounter each other.
- Short complementary stretches of duplex DNA form.
- The DNA then zips back together to form the original structure.

Obviously if a single molecule is denatured it should be able to reform completely. But if two separate molecules are denatured together, for example wheat and barley DNA, then complementary regions between the two DNAs will be able to form duplex DNA. The ability of two molecules to renature is called hybridization.

Hybridization: The pairing of complementary nucleic acids. Performing hybridization experiments in a solution is called liquid hybridization. We have already discussed the related procedure called filter hybridization.

COMPARATIVE GENOME MAPPING

Often the clones used to identify RFLPs in one species can be used in a second species. These heterologous probes can then be ued to develop a RFLP map in the second species. Once the two species have been mapped, the relative evolution of the two species can be compared. Comparative mapping can identify inversions, translocation and duplications that have occurred. Genetic factors can also be assessed by comparing map distances of genes with conserved gene order in the two species.

A comparison of the related grass species sorghum and maize was made using maize clones. The diploid chromosome number of the two species is twelve. Many of the sorghum chromosomes contained regions from two of the maize chromosomes. This result may represent the ancestral duplication of chromosomal material. Furthermore, during maize evolution duplicate genes have also occurred to a greater extent than that seen in sorghum. Only nine inversions of gene order have been observed between the species. Maize and sorghum were next compared with regard to genetic distance. Conserved gene orders were compared, and these linkages measured 862 cM in maize and 835 cM in sorghum. Therefore, it can be concluded that since the divergence of sorghum and maize, large chromosomal changes have not occurred and much of the recombination distance has been maintained. The same type of analysis was performed with two more distant species, tomato and pepper. Little linkage conservation appears to have been maintained since the divergence of these two species, even though the chromosome number has been conserved. The largest conserved linkage block is a 63 cM block of tomato chromosome two that was located on chromosome I of pepper. Some chromosomes of pepper contained six distinct regions of the tomato genome (pepper chromosome X). These chromosomal breakages and rearrangements are not centromeric in nature, suggesting that evolution involved breakage throughout the genome. Because duplicated regions were found in pepper, it was concluded that the gene duplications events within pepper occurred after divergence from tomato. Linkage distances of conserved gene orders were quite similar.

PHYSICAL AND GENETIC DISTANCES

All distances found on a linkage maps are the product of genetic recombination and therefore are considered to be genetic distances. Two pairs of loci that are genetically the same distance apart may not be physically the same distance apart because of suppressed recombina-tion in the region between two of the loci. Those loci where suppression of recombination does occur will physically be farther apart. RFLP hybridizations to large fragments of DNA are necessary to accurately gauge the physical distance.These experiments are similar to other Southern hybridizations.

The DNA is cut with restriction enzymes first. The choice of enzymes is important because you want to generate large fragments. For these experiments you want to cut the DNA with enzymes that recognize 8 nucleotide sequences in the target DNA. Because these sites will be rare (every 65,536 bases on average) the digestion products will range from several hundred kilobase to several megabases. Because the normal gel systems do not separate large fragments very easily, it is necessary to run pulsed-field-gel-electrophoresis. This technique was first applied to the separation of yeast chromosomes and was quickly applied to large-scale mapping experiments.

Physical and genetic distances are then resolved by hybridizing clones which define closely linked genetic markers, to DNA that has been cut with rare cutting enzymes. What you are searching for is co-hybridization of the two clones to the same restriction fragment. If two clones hybridize to the same fragment, then the maximum distance between those two clones is the size of the restriction fragment. Because you already know the genetic distance between the loci, you can correlate the genetic and physical distances in this region. The following table give the relationship between the physical and genetic distance in three species.

Species	*Kilobases/centimorgan*
Arabidopsis	139
Tomato	510
Corn	2140

But remember, these values are estimates of a single region of the genome of each species, and it is quite common to see that two regions of the same species have different amount of DNA per genetic distance. Correlations between physical and genetic distances have also been performed using deletion stocks in wheat.

A series of deletions stocks of a specific wheat chromosome were analysed with a probes known to hybridize to that chromosome. "Each locus was assigned to the chormosome region between the breakpoint of the largest deletion where the band was present and the next larger deletion where the band was absent." (Werner *et al.* PNAS 89:11307–11311) The authors noted large discprencies between the physical and genetic maps of chromosomes 7B and 7D. Two loci that are located near the centromere of 7B are 7 cM apart genetically, yet the distance between the two loci spans 25% of the chromosome. Another region at the distal region of long arm of 7B which accounts for about 15 % of the chromosome is 91 cM long genetically. This points out the difference of recombination that can occur within a single chromosome.

LINKAGE DRAG

One goal of plant breeding is to introduce a gene from a donor parent to improve a cultivar for a specific trait. For example, a wild germplasm could be used as a source of disease resistance. At the same time the breeder does not want to carry any of the other genes from the wild germplasm that might reduce the a agronomic fitness of the cultivar.

The backcross method of plant breeding is one manner in which the introduction of a specific gene is accomplished. One genetic feature though of backcross breeding is linkage drag. This refers to the reduction in fitness in a cultivar due to deleterious genes introduced along with the beneficial gene during backcrossing. Molecular makers offers a tool in which the amount of wild or alien DNA can be monitored during each backcross generation.

First, though, lets look at the amount of alien DNA that can be maintained after a backcrossing programme. The accompanying figure handed out in class shows the chromosomal region around the Tm-22 allele in a number of tomato cultivars. This allele was introduced from the wild tomato species L. peruvianum, and it provides resistance to tobacco mosaic virus. Large variation in the amount of introgressed DNA was observed. For example, Craigella-Tm-22 contains 51 cM of wild DNA, whereas Vendor-Tm-22 and Nova-Tm- 22 each contain about 8 cM of introgressed DNA.

Backcrossed lines could be developed rapidly in which only a small amount of foreign DNA is linked to the gene of interest. After the first backcross all lines with the gene of interest could be screened with an RFLP that is 1 cM away. Those lines in which a crossover occurred at this marker would be selected. Then those lines would be crossed to the recurrent parent.

The progeny from these crosses would then be scored for a second marker 1 cM away on the other side of the gene. Again crossovers progeny could be selected. Thus, in only two generations the amount of wild DNA could be reduced to 2 cM. The key to this procedure is to have markers that are polymorphic between the two parents that are closely linked to the gene of interest. Young and Tanklsley applied these principles to the reduction of L. peruvianum DNA in Craigella-Tm-22. The authors were able to reduce the amount of DNA on one side of the Tm-22 allele from 47 to 7 cM in one generation by selecting for crossovers at the CD32A locus.

QUANTITATIVE TRAIT LOCUS

Traditional quantitative genetic research defined a quantitative trait in terms of variances. The total phenotypic was first partitioned into genetic and environmental variances. The genetic variance could then be further divided into additive, dominance and epistatic effects. From this information it was then possible to estimate the heritability of the trait and predict the response of the trait to selection. It was also possible to estimate the minimum number of genes which controlled the trait.

Mapping markers linked to QTLs identifies regions of the genome that may contain genes involved in the expression of the quantitative trait. But what functions could these genes be encoding. To answer this question we should consider a trait such as yield. What types of qualitative genes (genes inherited as simple genetic factors) could be involved in the expression of yield? The first event required for yield is meiosis.

Therefore any gene that is involved in gamete formation could potentially be considered a QTL. Any of the genes involved in the protein and carbohydrate biosynthetic pathways could also affect the final yield of a plant and could also be considered to be QTLs. As we saw above, the markers associated with a QTL each account for only a portion of the genetic variance. Likewise each of these genes of known function may only account for a portion of the final yield. An important question that can now be posed is whether any known genes map as QTLs.

Beavis *et al.* analysed four populations of maize and found molecular markers linked to plant height. No marker was consistently associated as a QTL with plant height in all four populations. Each of the ten maize chromosomes contained a marker linked to a QTL for at least one of the four populations. The authors further were able to demonstrate that a number of the QTLs identified by the molecular markers mapped to regions containing genes known to have a qualitative effect on plant height.

For example, on chromosome 9 the gene d3 resides within 10 cM of a plant height QTL. This gene is involved in gibberellic acid sensitivity. Mutants do not respond to the hormone and do not undergo the normal cell elongation. These mutants are phenotypically shorter than normal maize plants. The question that needs to be raised is if the QTL that was being identified by the molecular marker is actually the d3 gene. It could be possible that what is actually being measured by the marker is the linkage of the marker with the gene. The statistical analysis of quantitative traits provided valuable information for the plant breeder. Molecular analysis of quantitative traits now provides new tools, not only as selection tools for plant breeding, but as starting points for the cloning of these genes. These objectives could not have been realized without molecular markers.

ANALYSIS OF GENOMES BY REASSOCIATION EXPERIMENTS

If a DNA molecule is melted and allowed to reassociate, the complexity of the genome dictates the rate in which duplex DNA will form. If we consider a simple molecule that consists of alternating GCs, this molecule will be able to form a duplex quicker than a molecule that consists of repeating blocks of AGCT. As the number of different combinations of bases increases, the time required for complete duplex formation to occur will increase.

Renaturation, or duplex formation requires random collisions between two single-stranded molecules. This process follows second-order kinetics and is concentration dependent. We will not go through the derivation of the formula but the important parameter used to define a certain DNA is:Cot½.

This value is defined as the amount of time required for one-half of the DNA to reanneal or form duplex DNA. The units for this parameter is moles of nucleotides per litre per second. The more complex the genome of interest, the longer it will take for like sequences to reanneal. Consequently, the Cot½ will be larger. Thus in terms of reassociation kinetics complexity has a specific definition.

Complexity: The total length of different sequences. For example, E. coli is considered to have a complexity of 4.2 × 106 base pairs. What is the experimental procedure used to derive these values?

In general the procedure is:

- Shear the DNA to be analysed to a length of about 300 bp.

- Melt the DNA (usually in 0.12 M phosphate buffer) by boiling for 5 min.
- Quickly place at 60°C.
- Take aliquots at different time points. Separate single-stranded DNA from double-stranded DNA by hydroxyapatite. Measure the amount of DNA that is double-stranded by absorbance at 260 nm.
- Plot the amount that is single-stranded versus the Cot value. The Cot value is expressed in log equivalent. This plot depicts the Cot curve.

When this type of experiment is performed with eukaryotic DNA three components are usually seen. These components each reanneal with their own unique Cot½ value.

The three components are termed the fast, intermediate, and slow components. Why do we see these three components? Eukaryotic genomes are characterized by sequences that are represented by different copy numbers. If a sequence is found many times in the genome, it will reanneal much quicker than those sequences that are found only once in the same genome. Thus the equivalent Cot curve for a eukaryotic genome will be different than a genome, such as E. coli, which only contains single copy sequences.

A comparison of the Cot value of each of these components with an E. coli standard allows us to derive the complexity of each component. The complexity of the slow component of the genome is greater than that for the other two components and is considered to represent the single copy portion of the genome. The complexity of the slow component can be used as a good estimate of the genome size.

The genome size will be the sum of the lengths of all the unique sequences. Using the example from Genes V - Lewin, p.664, the complexity of the slow component is 3×10^8 bp and the complexity of the intermediate component is 6×10^5. If we divide the complexity of the slow component into the intermediate component we get 2×10^{-3}. This demonstrates that the intermediate component contributes very little to the complexity of the genome. Therefore, the complexity of the slow or single copy portion of the genome can be considered equal to the genome size.

To derive the complexity of each component, it is necessary to run a standard, such as E. coli DNA, with each experiment. E. coli is considered to consist of only single- copy sequences. Let's say that in the experiment from which the Cot for each component was derived, the Cot½ value for E. coli was 4. Experimentally it was determined that the slow component comprised 45% of the total DNA. Therefore, if only that component was annealed, the Cot½ value would be 283 (630×0.45). That value is 71 (283/4) times slower than for E. coli. Therefore the complexity of the slow component is 71 times that of E. coli or 3.0×10^8 ($71 \times 4.2 \times 10^6$). The complexity of the other components is derived similarly. Genome size is quite variable throughout the biological world and the genome size in plants shows the greatest variation of any kingdom in the biological world.

Table. Variation in Genome Size among Plants

Species	*kb/haploid*	*pg/haploid*
Arabidopsis	7×10^4	0.15
Lily	1×10^8	100.00

Conversion factor: 1 pg = 0.965 × 109 bp = 6.1×10^{11} daltons

One manner in which a genome can be described is by determining the distribution of fast, intermediate and slow components in the genome. For comparison purposes, what does the human genome look like? Distribution Sizes Among Components of the Human Genome

Component	*% of Genome*
Fast	6
Intermediate	38
Slow	50

The table Sequence Distribution of Selected Plant Species lists the different components from different plant species. As you can see, plant species exhibit a wide range of values for each of the components. The genome of Arabidopsis is essentially entirely single copy sequences (the repetitive sequences have been determined to be essentially all chloroplast DNA).

At the other extreme, pea and wheat genomes have only 10–20% single copy sequences. Reassociation kinetic experiments of polyploid species, such as bread wheat (Triticum aestivium) were unable to derive a component that displayed true single copy kinetics. Instead the slowest component appeared to act as if it consisted of copies represented three times.

This result is consistent with the current hypothesis that bread wheat was developed from the introgression of three diploid wheat species. Genomic analysis suggest that this hypothesis is correct since the slowest component appears to consist of sequences represented three times.

REPEATED SEQUENCES

The intermediate and fast components are composed of sequences that are found many times in the genome. These sequences are called repetitive sequences and can vary in size from a 100 bp to 1000 bp or more. Furthermore, these sequences have undergone sequence divergence by the addition or deletion of sequences or by changes in the base pair sequence. Thus, the repeated sequences themselves show some divergence. An example of a highly repetitive sequence is the repeat found to be associated with the knob heterochromatin of corn. It ranges from 3-5 × 10^5 copies on a small knob to 1 × 10 ^ 6 on the large knobs. This sequence is unique to knobs and is not found associated with any other heterochromatic regions of corn. An example of a functional repeated sequence in plants is the corn storage proteins, zeins. Two major classes of zeins exist, the 22 and 19 kd classes. Sequence analysis has shown that both classes have the same structure. The only difference between the 22 and 19 kd class is the repeat unit. The 22 kd class has 8 repeat units and the 19 kd class has 7 repeat units. Estimates have been made of the number of copies of these genes and 30-50 copies of the 22 kd class are found in the corn genome. Thus, the repeat unit would be represented 240-400 times in the genome.

ORGANIZATION OF SINGLE-COPY SEQUENCES

Single-copy sequences are interspersed throughout the plant genome. These sequences are bounded by repeat sequences. The length of the single-copy regions varies widely among plant species.

In general, two types of arrangements are recognized:

- *Short Period Interspersion*: Single copy sequences of 300-1200 bp are interspersed as islands among short lengths of repeat sequences
- *Long Period Interspersion*: Single copy sequences of 2000-6000 bp are interspersed as islands among repeat sequences

How can the interspersion type be determined? Reassociation kinetic experiments are performed with 300 nt long fragments. These experiments give a characteristic Cot curve that defines each of the components. But what happens if the fragment that is followed is of a longer length, for example 900 nt? This fragment could be considered to consist of three 300 nt fragments, and each fragment may be from any of the three components.

Indeed this is how these experiments may be performed. A tracer of a longer length (such as 900 nt) is prepared and radiolabelled, for example with H^3. This tracer is added to a normal Cot reaction in such a low concentration that the rate is not affected. The reaction is then allowed to proceed to a Cot value determined from a normal experiment with only 300 nt fragments where the repetitive sequences have reannealed, but the single-copy sequences have not.

Then the amount of tracer which remains single-stranded is determined. What will be seen is a reduction in the amount of expected single-copy sequences. Why? If the 900 nt long fragment contains both single copy and repetitive sequences it will reanneal under these conditions as a repetitive sequence, and thus the amount of DNA that appears in the single-copy fraction will be reduced proportionally. Therefore, that single-copy sequence is interspersed with repetitive sequences. If for example the single copy fraction is determined to account for 40% of the genome when 300 nt fragments are used, and we calculate that the genome has 30% single-copy sequences when a 900 nt tracer is used, 10% (40%-30%) of the 900 nt fragments contain single-copy and repeat fragments and that 25% (10%/40%) of the single-copy DNA is interspersed with repetitive sequences at an interval of 300-900 nt. A final concern is the size of the repeat units. This can be obtained by using a long tracer (5000 nt, for example) and reannealing to an appropriate Cot such that only repeat sequences bind to the tracer. The product is then treated with an enzyme that cuts only single-stranded DNA. This enzyme is S1 nuclease. After digestion you obtain products that only contain repeat sequences. These products are then sized to give the average, mode and range of repeat sequences.

EVOLUTION OF REPEATED SEQUENCES IN CEREALS

The analysis of repeated sequences has also provided experimental evidence which supports the current model of cereal speciation. The following tables and the associated diagram detail the results. These show that once a repeat enters the lineage it remains. Therefore group I repeats are found in all of the species because it is the most ancient repeat. Secondly, new repeats are added to the lineage and appear in all species which subsequently diverged from the lineage. Finally, species-specific repeats are formed in each species after that species has diverged from the common lineage. The distribution of the lineage specific repeat supports the speciation model.

ESTIMATING THE NUMBER OF EXPRESSED GENES

Reassociation kinetics can also estimate the number and abundance of expressed genes. These experiments are performed using high concentrations of RNA and tracer levels of either DNA or cDNA.

These experiments are analagous to DNA reassociation experiments except the units are expressed as moles of ribonucleotides per litre per second and the value is called a Rot. If tracer amounts of DNA are allowed to reanneal only a portion of that DNA will form a duplex because not all the DNA will be expressed in the RNA population to which it is being hybridized.

Let's use the example in the Genes V to derive an estimate for the number of genes that are being expressed. In this experiment 1.35% of the DNA hybridized to the RNA. Since RNA is single stranded the other anti-sense strand of the DNA would not have a partner with which it could hybridize. Thus actually 2.7% of the DNA is represented in this RNA population. If the genome size is 8.1 X 108 bp and the single-copy sequences represent 75% of the genome then we can estimate the complexity of the expressed DNA.$0.027 \times 0.75 \times (8.1 \times 108$ bp) $= 1.7 \times 107$ bp If each gene is about 2000 base pair then the number of genes that is expressed is:7×107 bp / 2000 = 8500 genes

CHLOROPLAST GENOME ORGANIZATION

All angiosperms and land plants have cpDNAs which range in size from 120-160 kb; three expceptions are:

Species	Size (kb)
N. accuminati	171
Duckweed	180
Geranium	217

All cpDNA molecules are circular and spinach is used as the basis for all comparisons. Very few repeat elements are found other than short sequences of less than 100 bp. The notable exception is a large (10-76 kb) inverted repeat section, which when present, always contains the rRNA genes. (Legumes such as pea do not contain this repeat.) For the majority of species, this repeat region is 22-26 kb in size. Finally,the genetic order of the ribosomal unit is conserved in all species:

$$16S - tRNA^{ile} - tRNA^{ala} - 23S - 5S$$

Two other features of chloroplast DNA is described. First it was shown to that it can exist in in two orientations This implies that the molecule can undergo an isomerization event. Second is has been shown that spinach, corn, tomato and pea can all exist as multimers (PNAS 86:4156, June 1989).

Multimer	*Relative Abundance*	*Per cent*
Monomer	1	67.5
Dimer	1/3	22.5
Trimer	1/9	7.5
Tetramer	1/27	2.5

Because photosysnthesis is the primary function of the chloroplast it is not surprising that the chlroplast genome contains genes which encode for proteins that are involved in that process.

Reaction	Function
Dark Reactions	rbcS (nuclear encoded)
	rbcL (chloroplast encoded)
Light Reactions	apoproteins for PSI andPSII
	cytochrome b6
	cytochrome f
	6 of 9 ATPase subunits
	cab, LHC proteins (nuclear encoded)
	plastocyanin (nuclear encoded)
	ferredoxin (nuclear encoded)
Other	19/60 ribosome binding proteins
	translation factors
	RNA polymerase subunits
	tRNA and rRNA genes

Atrazine resistance is apparantley mediated through the psbA gene sequences of the 32 kd protein which is encoded by cpDNA. DNA sequence analysis revealed the following amino acid changes that are thought to be important.

Species	AA#	Susceptible	Resistant
Blue green algae	264	Ser (TCG)	Ala (GCG)
Chlamydomonas	264	Ser (TCT)	Ala (GCT)

Solanum nigrum	264	Ser (AGT)	Gly (GGT)
Amaranthus	228	Ser (AGT)	Gly (GGT)

Evolutionary Changes of cpDNA

- The majority of changes are small insertions and deletions of 1-106bp; significantly, a few length mutations of 50-1200 bp are clusted in "hot spots".
- The largest deletion occured in pea where an entire rRNA cluster is lost.
- The most common evolutionary change is in gene order. Small changes in the gene order occur, especially in the algae, but inversions have generated large scale order changes:
  - *Legumes*: About 50 kb inversion brought rbcL closer to psbA
  - *Wheat*: About 25 kb inversion brought atpA closer to rbcL

MITOCHONDRIAL GENOME ORGANIZATION

In comparison to the chloroplast genome, the size of the mitochondrial genome is quite variable.

Species	*Size (kb)*
Oenothera	195
Turnip	218
Corn	570
Muskmelon	**2400**

Further, in comparison to the mitochondrial genomes of other species the size is quite large and variable.

For example, animal mitochondrial genomes range in size form 15-18 kb, and fungi mitochondrial genomes range form 18-78 kb. Plants may code for more proteins than with species. For example, genes for ribosomes, subunits I and II of cytochrome oxidase and ATPase subunits are located on the mitochondrial genomes of plants.

When DNA from corn mitochondria was investigated with EM, several circular molecules of different sizes were detected. Once the genome was mapped it became apparent that a mechanism existed to generated these circles of different sizes. It is now understood how these molecules arise. First, lets look at the simple situation of turnip. Two direct repeats undergo intramolecular recombination to give the two smaller molecules:

218 kb † 135 kb + 83 kb

The mitochondrial genome of corn undergoes the same type of recombination, but the events are more complex. First, the master circles can be subdivided into two major subgroups:

570 kb † 488 kb + 82 kb

570 kb † 503 kb + 67 kb

The second group of molecules are still labile and can produce several other subpopulations. Further two subgenomic circles can unite to form a larger circle. This variability is possible because corn has 10 repeats with which intramolecular recombination can occur.

Species	Master Circle **Size (kb)**	Sub-genomic **Circle Size (kb)**	Repeat Size (kb)
Turnip	218	135 + 83	2
Cauliflower	217	172 + 45	?
Black Mustard	231	135 + 96	7
White Mustard	208	none	none
Radish	242	139 + 103	10
Spinach	327	234 + 93	6

Introns have been located in the cytochrome oxidase subunit II gene (the cytochrome complex consists of 3 mitochondrial and 4 nuclear encoded genes). This gene contains one intron in rye, corn, wheat, rice, and carrot, but for other species such as Oenothera, broad bean, cucumber the gene has no intron.

Promiscuous DNA

Stern and Lonsdale (1982) hybridized mtRNA to a SstII digest of maize mt DNA and found that it hybridized to fragments known not to contain mt rRNA genes. The question of interest was - what was it hybridizing to? They next looked at a cosmid clone of corn mtDNA that hybridized to the mt RNA and found that it hybridized to a RNA molecule of the size of the cp 16S RNA gene. How could this have happened?

They next mapped the clone and compared it to the map of the corn cpDNA and found that the clone map was almost congruent with that of the the cpDNA 16S RNA region. This mapping showed that the two maps were nearly identical over a 12 kb region of DNA. These results suggest that cpDNA had been transferred to the mitochondrial genome.

The observation that organelle DNA was found in other DNA compartments of the cell was extended by other researcher. Stern and Palmer looked at corn, mung bean, spinach and pea and found extensive evidence of cpDNA/mtDNA homology.

These observations were extended to other DNA locations in the plant cell. Kemble *et al* (1983) demonstrated that mitochondrial DNA sequences are located in the nucleus of corn. Scott and Timmis (1984) showed that cpDNA sequences are found in the nuclear DNA.

RNA EDITING

The Central Dogma of Molecular Genetics states that the information that is found in DNA is used to produce mRNA molecules that are instrumental in the production of proteins. Therefore, the information flows directly from DNA to protein, via the RNA intermediate molecule. Recently it has been discovered that the information that is contained in the DNA is not always found in the RNA products used to make proteins.It has now been demonstrated that mitochondria and chloroplast contain the biochemcial machinery to alter the sequence of the final transcription product. This process is called RNA editing. This process was identified in the following manner. Sequence analysis of a number of cytochrome c oxidase subunit II genes from non-plant species revealed that a tryptophan residue was invariant at several locations in the final protein product. But sequence analysis of this gene in several plant species revealed arginine at those positions.

Since a single base pair change in the codons for the two amino acids could generate this change (CGG for UGG), it was suggested that CGG encoded for tryptophan and not arginine in plant mitochondria. (This is the only change in codon usage that has been suggested for plants and has been postulated for several other genes as well.) But this change in codon usage was not universal, that is some CGG codons actually specified arginine in the final protein product. Furthermore, no amino acyl tRNA that recognized CGG was found to be charged with tryptophan, a prerequisite if this codon specification was actually real.

The solution to this dilemma was found by sequencing the mRNA products for cytochrome oxidase subunit II genes. It was found that in the mRNA the cytosine residue had been changed (edited) to uridine at the sequence location where the invariant tryptophan residue is found. This changed the codon at that location to UGG which is recognized by a tRNA that carries the amino acid tryptophan. An analysis of three other plant mitochondrial genes where the same altered codon usage was predicted suggested that mRNA editing was also occurring at the codon and that a cytosine residue was edited to uridine.

This editing process has also been detected in protozoa and it remains to be determined if RNA editing is a widespread function in mitochondria. A final point that this editing function highlights is that the sequence that is found in the DNA is not entirely and faithfully represented in the final protein product.

Specific Features of RNA Editing

- Editing can occur in both mitochondria and chlorplasts
- To date, >300 different editing events have been detected in plant mitochondria.
- The vast majority of the events involve a C to U tansition. A few cases of U to C transitions have been reported. This suggests that the editing machnery can also carry out the reverse modification.
- RNA editing can modify from 0.8% to 5.8% of the nucleotides of a specific transcript.
- The RNA editing events appear to occur at random in the transcript.
- Both 5' and 3' non-coding regions of mRNAs have also been shown to be edited.

- Structural RNAs such as tRNAs and rRNAs do appear to be affeceted.
- Editing can convert a tryptophan codon to a arginine codon (CGG to UGG).
- Start AUG codons can be created from ACG threonine codons
- Stop codons can be created by editing CAG, CAA and CGA codons.
- The most frequent amino acid substitions derived from RNA editing are Pro to Leu, Ser to Leu and Ser to Phe.
- Plant mitochondria do not use the universal genetic code.

The primary benefit of RNA editing could be evolutionary conservation of protein structure. For example, bound copper is required for the funciton of cytochrome c oxidase subunit II (coxII). After editing, all amino acids #228 are converted to cysteine, an amino acid required for copper to bind. In all species except for plants, the coxII gene encodes for methionine at codon #235. In plants, this methionine is generated by RNA editing. These events suggest that this protein is under very strong structural and functional constraints.

FUNCTIONAL GENOMICS IN PLANT CELL BIOLOGY

Genomic resources have significantly impacted plant biology research in recent years. Cell biology has been further enabled by an ongoing revolution in visualization technologies. Using fluorescent proteins (FPs), we now have unprecedented views of cellular architecture, and we can study real-time dynamics of cell structure, function, and protein localization. To date, these technologies have been most widely used in Arabidopsis (Arabidopsis thaliana); however, the grasses provide a unique opportunity to study the underlying mechanisms and inter-related controls of cell growth, morphogenesis, and physiology in leading crop models.

Plants are photoautotrophs: The most obvious distinguishing characteristic of plants is the presence of chloroplasts, which enable them to photosynthesize. They are thus photoautotrophs. The presence of plastids also has implications for genetics, since plastid genomes can carry important traits, most of which are directly related to photosynthesis.

Cell walls constrain development: Plants have cell walls made of cellulose, fungi have chitinous cell walls, while animals do not have cell walls. Plants maintain their structural integrity through a combination of mechanisms: turgor, cell walls, and insoluble components such as lignin. Animals derive their structural integrity through endo- or exo-skeletons, muscle and connective tissue. Higher plants and animals have circulatory systems that allow for the distribution of nutrients and the removal of wastes. Fungi are constrained in size due to the lack of any comparable features.

- Another consequence of the cell wall has to do with the limitations it imposes on diffusion. With pores of 3.5 to 5.2nm, movement of molecules >15Kd is significantly impeded. Thus, plants must subsist on molecules of low molecular weight, and any intercellular signaling molecules that have to pass through the cell wall must also be small.

- *Plasmodesmata facilitate symplasty*: Plasmodesmata convert a plant from being just a collection of individual cells to a large, interconnected symplast. Although the passage of molecules through plasmodesmata is tightly controlled, this provides a means for intercellular communication that is not available in animal tissues.
- *Most cell volume is taken up by vacuoles*: Often 50%, to as much as 95% of the volume of a plant cell is taken up by vacuoles. The vacoule makes it possible for plant cells to become very large, without a correspondingly large increase in cytoplasmic volume. This is important because the cytoplasm is requires a fairly large investment, in terms of macromolecular synthesis and turnover. Additionally, by limiting the volume of the cytoplasm, the rates of intracellular diffusion are not slowed down. This in turn affects diffusion-dependent processes, such as signal transduction from the plasmalemma or cytoplasm to the nucleus, protein synthesis, protein transport, membrane synthesis, cytoplasmic streaming, and absorption and equilibration of nutrients, gases, and other small molecules.

LIFE HISTORY OF PLANTS

- *The meristem (germline)*: The germ line - Animals have specialized germline tissues, spermatogonia and oogonia, which diverge from somatic cells early in embryogenesis, and are largely insulated from selective pressures. In contrast, flowering plants have large portions of their undifferentiated tissue as meristem, tissue capable of growing vegetatively and differentiating (often in response to very specific environmental cues, such as day length) into reproductive organs which then produce germ cells.
- *Gametophyte and sporophyte generations*: In animals, the final meiotic products (spermatids and ova) are mostly metaboli-cally dormant, and actually shut down gene expression when they have completed maturation. In fact, egg cells in animals typically have enough mRNA and rRNA pre-synthesized to last the zygote for several cell divisions after fertilization.

Flowering plants have a 2 part life cycle, a diploid (or greater) sporophyte stage, and a haploid gametophyte stage.

Female: Megaspore mother cell (2N) † megaspore (1N) † megagametophyte † egg

Male: Microspore mother cell (2N) † microspore (1N) † pollen grain † sperm

Thus plant germline (gametophyte) cells undergo a number of cell divisions in the haploid state, and, presumably, have to perform all of the normal metabolic housekeeping functions. An important implication of this is that recessive lethal alleles in animals will be carried into the zygote, and selection against these traits may be prevented by heterozygosity. In plants, a good number of recessive lethals (eg. metabolic pathways, cell cycle controls) would be weeded out during the haploid gametophyte generation. One piece of evidence that could corroborate this is that it is possible to produce haploid plants from some species (eg. canola, tobacco). Admittedly, these plants are usually pretty sickly, but the fact that you can do this at all may be a result of this weeding out of deleterious recessive alleles.

- *Totipotency*: Plant somatic cells remain totipotent throughout the life of the plant. Animal cells tend to become terminally differentiated early in development, and can undergo genetic rearrangements which destroy totipotency. Examples: selective loss of chromosomes from non-germline cells in some invertebrates; selective polytenization of parts of chromosomes; somatic amplification of rRNA genes; rearrangements within immunoglobulin genes in lymphocytes.

*To elaborate on this last exampl*e: In vertebrates, the immunoglobulin genes in lymphocytes undergo rearrangements of DNA in the regions encoding the hypervariable regions of antibody proteins. Since a different rearrangement occurs in each cell, the population of lymphocytes can differentiate into T-cells and B-cells, each of which will carry antibodies with unique specificties. These DNA rearrangements are irreversible. The totipotency of plant cells has important implications. First, it deprives plants of one, admittedly drastic, means of genetic regulation. More importantly, it has made the transformation of plants for experimental and genetic engineering purposes a great deal easier.

- *Differentiation without migration*: Animal cells must migrate by amoeboid motion through the embryo during development. Plant cells are constrained by their cell walls to stay where they are. Therefore, differentiation must be carried out without migration. This implies that there are very fundamental differences between plant and animal ontogeny. Is this inability of cells to migrate an evolutionary dead end for plants? Is this the key to why animals seem to have a greater potential for physiological complexity than plants?
- *Continuous programme of development*: The totipotency of plant cells, along with the relative lack of terminal differentiation during vegetative growth results in a continuous programme of differentiation and development throughout the lifetime of the plant. In animals, once an organ or tissue has formed, that's it. You could say that, while animals go through a single iteration of the developmental programme, plants can go through multiple iterations, usually in response to environmental stimuli. This can be viewed in its extreme as the abilty of new plants to be produced by vegetative propogation (eg. potatoes) without ever going through a gametophyte stage.

- *Shift between vegetative and reproductive growth*: In animals, germline cells become differentiated very early in development, and specialized sex organs are established early. In flowering plants, sporophytes grow vetatively for an extended period. The shift from vegetative to reproductive growth is triggered by environmental cues. It is during flowering that reproductive organs develop.
- *Kinetics of cell growth*: The same kinetic principles that govern bacterial cultures govern plant growth. In a sense, cells in an organ can be thought of as a culture.

Per centage growth per day as a function of age: The growth of a local population of cells is limited in part by the ratio of the mass of cells to vascular tissue.

Hypothesis: All the genetic programme needs to do is control the key points at which vascular tissue branches out, and the adjoining mesophyll and epidermal cells come along for the ride, filling out the shape, as it were.

Thus, things like the thickness of leaves, or the distance between veins are determined by diffusion limitations. Superficially, we're not just talking about growth as cell division, but there is also a component of elongation. Both these processes must be under the control of developmental regulators.

- *Developmental plasticity*: The plant body plan is more of an iterative loop, rather than a specific blueprint, as in animals. The specific location, and number or organs, is often environmentally determined:
  - Switch from vegetative to reproductive growth (flowering) does not necessarily occur at a specific stage of growth
  - Number of leaves or nodes
  - Tissues such as roots can grow in direction of nutrients; leaves and stems can grow towards light.
- *Localized senescence*:
  - leaves senesce and abscise, while branches persist. (One of the few analogues in animals might be a snake shedding its skin.)
  - biomass from senescent tissue can be resorbed and transported to other tissues eg. seeds.
- *Cell division not required for new cell identities*:
  - Mesophyll cells can differentiate into tracheary elements through a breaking down of the cytoplasm and vacuole, and thickening of the cell walls.

GLOBAL ANALYSIS OF GENE EXPRESSION

The experiments of Goldberg and others on global analysis of gene expression provided what is, even today, the most comprehensive look at the role of gene expression in development. The key problem with R0t analysis, as done in these studies, is that nothing is learned about which genes or pathways are important in each organ or tissue studied. We can estimate the complexities of the RNA populations, and even learn, for example, that gene expression in leaves is very similar to that in petals, but quite dissimilar to that in roots. However, not knowing the identities of the genes involved limits the usefulness of these findings.

Most of the history of molecular biology has been characterized by the use of cloned probes, typically in Northern blots, to measure gene expression one gene at a time. This is precise and specific, but still has important limitations. First, the number of genes that can be examined is limited for practical reasons. You have to have a clone for every gene you wish to study. Secondly, it is necessary to decide which genes to look at. Unfortunately, choosing which genes to study biases the experiment. Important genes might be missed, and the role of relatively minor genes could be overestimated. One of the unifying themes in the field of genomics is that by working on a large scale, you get a more comprehensive picture of the subject. We will look at several large scale approaches towards finding important genes in biological processes.

Gene Arrays

EST studies begin to tell us about the expression of the genome, but all they really tell us is that a gene is expressed in the cells from which the ESTs were derived. Gene array experiments give a more precise picture of the 'transcriptome', the set of transcribed genes, in specific developmental stages, or in response to treatments.

- *Gene array technology*: Measures mRNA levels for thousands of genes.
  - Any cell or tissue type.
  - At any point in development.
  - In response to any stimulus.

Gene arrays consist of hundreds or thousands of cDNAs spotted onto microscope slides (microarray) or nylon filters (macroarray). cDNAs are chosen from EST collections, so the sequences, and usually the identities of genes in the array are known. In a gene array experiment, an mRNA population is isolated from cells. The population is labeled by synthesizing complementary cDNAs using reverse transcriptase and labeled nucleotides. The resulting cDNA population is then hybridized to the array.

- *Gene* x: Strongly expressed; high abundance transcript
- *Fene* y: moderately expressed; medium abundance transcript
- *Gene* z: weakly expressed; low abundance transcript

Each transcript base pairs with the complementary DNA for its corresponding gene on the array. Signal strength is proportional to the abundance of each mRNA *Example*: cDNA probes were made from Human fibroblasts treated with serum (Cy5-dUTP) or serum-deprived cells (Cy3-dUTP). 3 replicate arrays were scanned, and the same regions from the replicate arrays are shown. Serum inducible genes appear green, while serum suppressible genes appear red. 1 - protein disulfide isomerase-related protein P5; 2 - IL-8 precursor; 3 - EST AA057170; 4- vascular endoghelial growth factor.

- *Two general classes of data*: Gene array studies tend to generate two different types of data. Studies in which two or more conditions are compared at a time generate discrete state data. Often it is critical to follow the expression of a gene over time after a treatment. In timecourse experiments, the expression of each gene in response to two or more treatments is measured over time. For example, in the timecourse at right, the solid blue and red dashed curves might represent the expression levels for a gene in response to two different drugs.

There is a whole family of problems in normalization of data and controlling for components of experimental variation. To put things into perspective, if the experiment was repeated 3 times, the timecourse above represents 2 treatments × 6 times × 3 replicates = 36 probes hybridized to 36 duplicate arrays to generate the data. Although the data for each replicate are averaged, there is often a great deal of variation in the results, which can potentially negate any meaning. Therefore, extraordinary measures must be taken to minimize experimental variation at each step in the procedure, to minimize the overall variation. What we're ultimately trying to get from gene array experiments is expression patterns for each of the hundreds or thousands of genes in the array. By identifying genes whose expression patterns are similar, we can discover which groups of genes work in concert, in response to a given stimulus.

SEED DEVELOPMENT

During seed development, the embryo differentiates into two organ systems: the axis and the cotyledon. The axis contains root and shoot meristems that will give rise to the mature plant after seed germination. By contrast, the cotyledon is a terminally differentiated organ system that senesces after germination and is responsible for synthesizing andstoringfood reserves used by the germinating seedling.

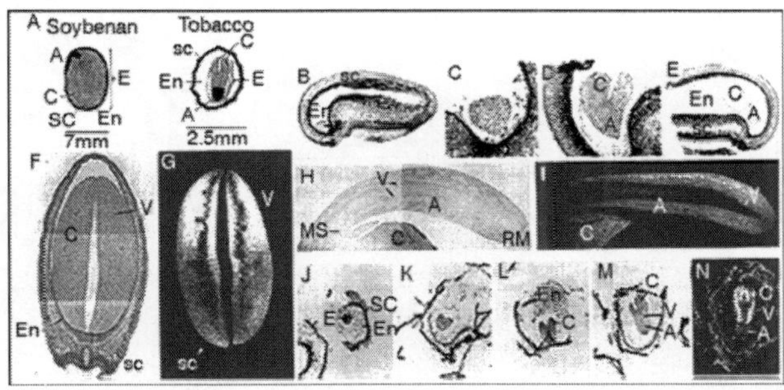

e#Conglycinin mRNA In Soybenan And Transgenic Tobacco.

e-conglycinin - storage protein, accumulates to high levels in soybean seeds. ~15 genes in soybean, clustered in two different chromosomal loci. 2-G - Dark-field microscopy shows silver grains as bright signal against dark background. e-conglycinin mRNA accumulates solely in the cotyledons (c) and not in surrounding embryonic tisssues (e). In the cotyledons, it accumulates in the storage parenchyma tissues and not in the vascular cylinder (v). 2J-M - development of tobacco seed. 2N - In-situ hybridization of transgenic tobacco expressing soybean beta-conglycinin.

- Similar expression patterns to soybean ie. only in cotyledon, non-vascular tissues
- Indicates that signals for tissue-specific expression in developing embryo must be well conserved.

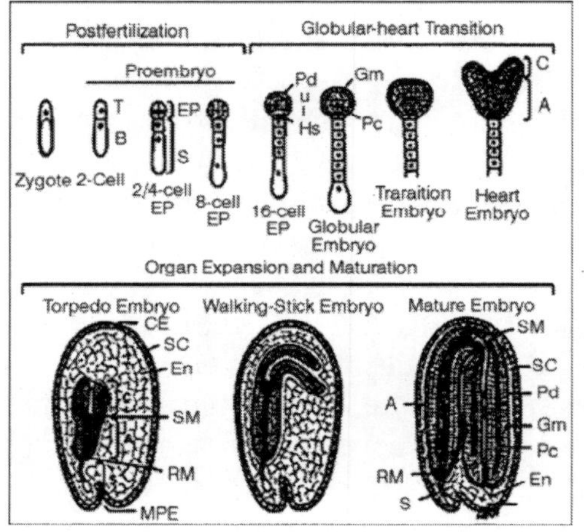

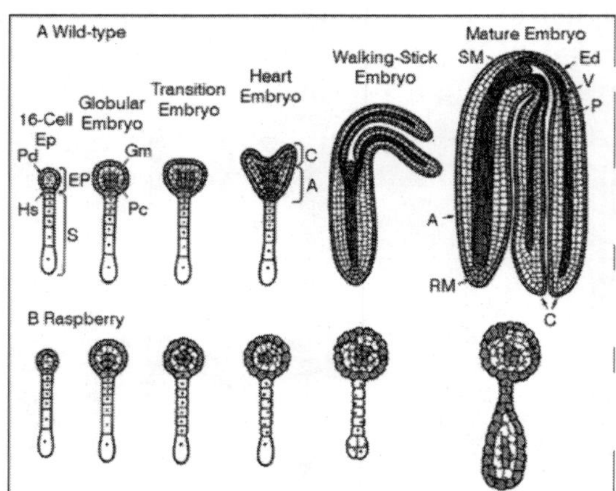

Fig. Overview of Embryogenesis

Embryogenesis

Plant embryogenesis is the process that produces a plant embryo from a fertilised ovule by asymmetric cell division and the differentiation of undifferentiated cells into tissues and organs. It occurs during seed development, when the single-celled zygote undergoes a programmed pattern of cell division resulting in a mature embryo. A similar process continues during the plant's life within the meristems of the stems and roots. Tobacco and soybean embryos contains only about 10 distinct cell and tissue types each, yet the mRNA populations contain roughly 20,000 different transcript classes. This is in contrast to animal development, in which a reduction in the number of genes expressed occurs as embryogenesis proceeds.

The Process of Embryogenesis

- Assymetric cleavage of the zygote results in the formation of a n embryo with a suspensor and embryo proper.
- *Embryo proper*: Gives rise to embryo.
- *Suspensor*: Anchors embryo to surrounding embryo sac and ovule tissue; serves as a conduit for nutrients from maternal sporophyte into developing proembryo. Suspensor senesces after the heart stage and is not a functional part of the mature seed.
- Embryonic organs and tissue-types differentiate during globular-heart transition phase.

Most of embryo is devoted to formation of the root. The hypophysis (Hs) gives rise to the root meristem.

Hypophysis: The uppermost cell of the suspensor from which part of the root and rootcap in the embryo are derived.

The shoot further differentiates into the axis and cotyledon. After the critical heart embryo stage, the shoot meristem is determined. The shoot meristem will give rise to the epicotyl, which in turn, will mature into the bulk of the plant body.

DNA FRAGMENT AND CLONING VECTORS

A cloning vector is a small piece of DNA into which a foreign DNA fragment can be inserted. The insertion of the fragment into the cloning vector is carried out by treating the vehicle and the foreign DNA with the same restriction enzyme, then ligating the fragments together. There are many types of cloning vectors. Genetically engineered plasmids and bacteriophages (such as phage o) are perhaps most commonly used for this purpose. Other types of cloning vectors include bacterial artificial chromosomes (BACs) and yeast artificial chromosomes (YACs).

Most commercial cloning vectors have key features that have made their use in molecular biology so widespread. In the case of expression vectors, the main purpose of these vehicles is the controlled expression of a particular gene inside a convenient host organism (eg. E. coli). Control of expression can be very important; it is usually desirable to insert the target DNA into a site that is under the control of a particular promoter. Some commonly used promoters are T7 promoters, lac promoters (bla promoter) and cauliflower mosaic virus's 35s promoter (for plant vectors).

To allow for convenient and favourable insertions, most cloning vectors have had nearly all their restriction sites engineered out of them and a synthetic multiple cloning site (MCS) inserted that contains many restriction sites. MCSs allow for insertions of DNA into the vector to be targeted and possibly directed in a chosen orientation. A selectable marker, such as an antibiotic resistance is often carried by the vector to allow the selection of positively transformed cells. All plasmids must carry a functional origin of replication.

Some other possible features present in cloning vectors are: vir genes for plant transformation, intergrase sites for chromosomal insertion, lacZd fragment for d complementation and blue-white selection, and/or reporter genes in frame with and flanking the MCS to facilitate the production of recombinant proteins [eg. fused to the Green fluorescent protein (GFP) or to the glutathione S-transferase.

TYPES OF VECTORS

- Plasmids.
- Bacteriophage.
- Cosmids.
- Yeast artificial chromosomes (YACs).
- Bacterial artificial chromosomes (BACs).

PLASMIDS

Characteristics of Plasmids

- Extrachromosomal circular DNA molecules which are not part of the bacterial genome.
- Size range: 1-200 kb.
- carry functions advantageous to the host such as.
  - Produce enzymes which degrade antibiotics or heavy metals
  - Produce restriction and modifying enzymes
- Replication is coupled to host replication in a:
  - Stringent manner: One (or two) plasmids made during each round of bacterial replication
  - Relaxed manner: 10-200 copies of the plasmid made during each round of bacterial replication; (this can be increased to 1000-2000 plasmids by stopping host protein synthesis and replication with the antibiotic chloramphenicol)

Foreign DNA is inserted into a plasmid (or any cloning vector) by ligating the DNA into a complementary site in the plasmid. These sites are generated by digesting the DNA and vector with the same restriction enzyme. (The site for the restriction enzyme that is chosen should only be represented once in the plasmid. Thus, when the plasmid is digested, a single, linear molecule would be generated.) The foreign DNA is then inserted into the plasmid by the action of the enzyme DNA ligase. The next step is to insert the ligated DNA into a bacterial cell for propagation. This is done by a technique called transformation. Bacterial cells are treated with either Ca2Cl or Rb2Cl. This treatment produces pores in the bacterial cell wall and membrane through which the plasmid enters. Although there is no size limitation to the ligation reaction, transformation efficiency is dictated by the size of the plasmid.

Process by which a plasmid is used to import recombinant DNA into a host cell for cloning. Many diseases are caused by gene alterations. Our understanding of genetic diseases was greatly increased by information gained from DNA cloning. In DNA cloning, a DNA fragment that contains a gene of interest is inserted into a cloning vector or plasmid.

The plasmid carrying genes for antibiotic resistance, and a DNA strand, which contains the gene of interest, are both cut with the same restriction endonuclease. The plasmid is opened up and the gene is freed from its parent DNA strand.

They have complementary "sticky ends." The opened plasmid and the freed gene are mixed with DNA ligase, which reforms the two pieces as recombinant DNA Plasmids + copies of the DNA fragment produce quantities of recombinant DNA. This recombinant DNA stew is allowed to transform a bacterial culture, which is then exposed to antibiotics. All the cells except those which have been encoded by the plasmid DNA recombinant are killed, leaving a cell culture containing the desired recombinant DNA. DNA cloning allows a copy of any specific part of a DNA (or RNA) sequence to be selected among many others and produced in an unlimited amount. This technique is the first stage of most of the genetic engineering experiments: production of DNA libraries, PCR, DNA sequencing, *et al*.

Inserting a DNA Sample into a Plasmid

Plasmids are similar to viruses, but lack a protein coat and cannot move from cell to cell in the same fashion as a virus. Plasmid vectors are small circular molecules of double stranded DNA derived from natural plasmids that occur in bacterial cells. A piece of DNA can be inserted into a plasmid if both the circular plasmid and the source of DNA have recognition sites for the same restriction endonuclease.

Restriction enzymes, also called restriction nucleases (EcoRI in this example), surrounds the DNA molecule at the point it seeks(sequence GAATTC). It cuts one strand of the DNA double helix at one point and the second strand at a different, complementary point (between the G and the A base). The separated pieces have single stranded "sticky-ends," which allow the complementary pieces to combine. The newly joined pieces are stabilized by DNA ligase. EcoRI, one of many restriction enzymes, is obtained from the bacteria Escherichia coli.

The plasmid and the foreign DNA are cut by this restriction endonuclease (EcoRI in this example) producing intermediates with sticky and complementary ends. Those two intermediates recombine by base-pairing and are linked by the action of DNA ligase. A new plasmid containing the foreign DNA as an insert is obtained. A few mismatches occur, producing an undesirable recombinant. The new plasmid can be introduced into bacterial cells that can produce many copies of the inserted DNA. This technique is called DNA cloning.

Effect of Plasmid Size on Transformation Efficiency

Molecule size (kb)	% Maximum probability
2.0	57
3.2	100
4.3	86
12.5	43
20.0	36
39.0	14
54.0	6

The goal of the ligation reaction is to insert the foreign DNA into the vector. An unwanted ligation product also occurs - religated vector. One way to minimize this event is to treat the vector with phosphatase. This removes the terminal phosphate group from the restriction site of the vector and theoretically prevents religation of the two ends. In reality though, this treatment is never 100%, so a low level of religation does occur. Thus after transformation you will have two types of cells: those which contain the original plasmid and those that contain a plasmid containing foreign DNA.

Plasmids are designed to distinguish the two types of transformation products. pBR322, the first widely used vector, utilizes differential antibiotic screening to distinguish the two types of transformation products. Let's say that we clone into the BamHI site of the vector.

The insert DNA will then split the gene responsible for tetracycline resistance. But at the same time, the gene for ampicillin resistance is left intact. Transformed cells are first grown on bacterial plates containing ampicillin. This will kill all the cells that do not contain a plasmid. But we still cannot say which cells contain foreign DNA. Those cells that grew on ampicillin are then replica plated on plates with ampicillin and tetracycline. Those cells which grow in the presence of the ampicillin, but die under tetracycline selection contain plasmids which have foreign DNA inserts.

pBR322 was a breakthrough for molecular biology, but the double screening procedure was time consuming and could be subject to error. In 1981, a new series of plasmids were developed that permitted the identification of the foreign DNA containing cells in a single screening step. These are called the pUC plasmids. As with pBR322, ampicillin resistance is used as one selectable marker. The second marker is based on insertional inactivation of the E. coli lacZ gene. The wild type gene can hydrolyze a specific dye [X-Gal (5-bromo-4-indoyl-B-D- galactopyranoside)] to a blue colour, and the bacterial colony is stained blue.

A multiple cloning site has been inserted into this gene. This site will accept fragments ending in a number of different restriction enzymes. Upon insertion of DNA into this site, the activity of the gene is eliminated and the colony appears white in colour. Thus transformed colonies containing plasmids with inserts can be distinguished from those with plasmids without inserts based on the colour of the colony and the ability of the colony to grow on an ampicillin containing media.

cDNA Cloning (Cloning Eukaryotic mRNA)

cDNA cloning is a method of obtaining a DNA copy of the mRNAs that are expressed at a specific stage in the development of the plant. In this manner, you can enrich the library that you will screen for those sequences in which you are interested. The reagent required for this type of cloning approach is mRNA. These mRNAs have a poly A+ tail.

This tail permits the isolation of poly A+ mRNA by using either oligo-dT or oligo-U columns. Total RNA is run through one of these columns under conditions which favour the binding of the tail to the matrix on the column. After the column is extensively washed, the conditions are changed and the bound mRNA is isolated. This is the starting reagent for cDNA cloning.

Steps in cDNA Cloning

- Bind oligo-dT to the poly +A tail of the mRNA.
- Add reverse transcriptase and make a DNA copy of mRNA (cDNA). Erase the mRNA with alkali and high temperature. (First strand synthesis)
- Add a C-tail to the 3' end of the cDNA with terminal transferase.
- Add oligo-dG to the tailed-cDNA and make the second strand with reverse transcriptase or the Klenow fragment of DNA Polymerase I.
- Add dC's to the 3' end of the double-stranded cDNA with terminal transferase
- Add C-tailed, ds-cDNA to G-tailed, PstI cut pBR322 and anneal. (DNA ligase is not required for this step.)
- Transform E. coli cells.
- Select Tetr/Amps cells. Most, but not all inserts can be removed by PstI digestions.

BACTERIOPHAGE LAMBDA VECTORS

Extensive research has been directed towards the development of multipurpose lambda vectors for cloning ever since the potential of using coliphage lambda as a cloning vector was recognized in the late 1970s. An understanding of the intrinsic molecular organization and of the genetic events which determine lysis or lysogeny in lambda has allowed investigators to modify it to suit the specific requirements of gene manipulations. Unwanted restriction sites have been altered and arranged together into suitable polylinkers. The development of a highly efficient in vitro packaging system has permitted the introduction of chimeric molecules into hosts. Biological containment of recombinants has been achieved by introducing amber mutations into the lambda genome and by using specific amber suppressor hosts. Taking advantage of the limited range of genome size (78 to 105% of the wild-type size) for its efficient packaging, an array of vectors has been devised to accommodate inserts of a wide size range, the limit being 24 kbp in Charon 40.

The central dispensable fragment of the lambda genome can be replaced by a fragment of heterologous DNA, leading to the construction of replacement vectors such as Charon and EMBL. Alternatively, small DNA fragments can be inserted without removing the dispensable region of the lambda genome, as in lambda gt10 and lambda gt11 vectors. In addition, the introduction of many other desirable properties, such as NotI and SfiI sites in polylinkers (*e.g.*, lambda gt22), T7 and T3 promoters for the in vitro transcription (*e.g.*, lambda DASH), and the mechanism for in vivo excision of the intact insert (*e.g.*, lambda ZAP), has facilitated both cloning and subsequent analysis. In most cases, the recombinants can be differentiated from the parental phages by their altered phenotype.

Libraries constructed in lambda vectors are screened easily with antibody or nucleic acid probes since several thousand clones can be plated on a single petri dish. Besides the availability of a wide range of lambda vectors, many related techniques such as rapid isolation of lambda DNA, a high efficiency of commercially available in vitro packaging extracts, and in vitro amplification of DNA via the polymerase chain reaction have collectively contributed to lambda's becoming one of the most powerful and popular tools for molecular cloning.

We have talked about plasmids as vectors for cloning small pieces of DNA. The limitation of this vector is the size of DNA that can be introduced into the cell by transformation. This presents problems when you are trying to create a genomic library of a large genome such as with plants. A genomic library contains all of the DNA found in the cell of the plant (or any organism). If you digest plant DNA to completion with a restriction enzyme, ligate those fragments into a plasmid vector and transform bacterial cells, only a portion of those fragments will be represented in the final transformation products. If a gene of interest is located on a large fragment then you will not be able to isolate that gene from a plasmid library. But what can be done to increase the probability of obtaining a clone which contains the entire gene. First you need to use a vector that can accept large fragments of DNA. Examples of these are bacteriophage and cosmid vectors and more recently yeast artificial chromosomes. Bacteriophage lambda vectors were developed because several observations were made that suggested that they could compl

ete their life cycles even if foreign DNA was inserted into a portion of its genome. This suggested that certain regions of the virus were not essential. Let's first discuss the life cycle of lambda.

- *Adsorption*: The phage particle binds at a maltose receptor site of the bacterial cell; growing the cell in the presence of the sugar increase the number of receptor sites
- *Penetration*: DNA is injected into the cell; at this point it can enter one of two pathways;
  - *Lysogenic pathway*: The phage DNA becomes integrated into the genome and is replicated along with the bacterial DNA; it remains integrated until it enters the lytic pathway
  - *Lytic pathway*: Large scale production of bacteriophage particles that eventually leads to the lysis of the cell; base pairing at the cos site leads toa circular molecule.

The linear double-stranded DNA lambda genome contains about 50,000 nucleotide pairs and encodes 50-60 different proteins. When the lambda DNA enters the cell the ends join to form a circular DNA molecule.

The bacteriophage can multiply in E. coli by a lytic pathway, which destroys the cell, or it can enter a latent prophage state. Damage to a cell carrying a lambda prophage induces the prophage to exit from the host chromosome and shift to lytic growth (green arrows). The entrance and exit of the lambda DNA from the bacterial chromosome are site-specific recombination events

- *Early transcription*: Transcription proceeds from the pL and pR promoters, through the N and cro genes and stops at terminators tL and tR1; a low level of transcription through the O and P genes occurs and terminates at tR2; the N product is an antitermination factor that is important for the next stage of transcription.
- *Delayed early transcription*: The N product binds to RNA polymerase and transcription proceeds past the tL, tR1 and tR2 terminators; genes to the left of N, involved in recombination, to the right of cro, involved in replication, are expressed at this point; another protein expressed from the Q gene is used for antitermination of later transcription.
- *Replication*: Early replication is through a theta form initiated from a single origin of replication site; later replication is via rolling circle replication; this produces long concatamers of the phage DNA that are cleaved at the cosL and cosR sites.

- *Late transcription*: The protein product of the cro gene builds up to a critical level and then binds to the oL and oR to stop early transcription; another protein, a product of the Q gene, has built up and activates transcription at the p'R promoter by antitermination; transcription terminates with in the b region; this transcription results in the production of the proteins required for the head and tail of the mature phage particle and those required for bacterial cell lysis.
- *Assembly*: A prophage head is produced; a unit length DNA is placed into the head by the action of the Nu1 and A proteins; the DNA is locked into place by the D protein and ter function of the A protein clips the DNA at the cosL and cosR sites; the concatamer is released, the tail is added and the mature phage particle is completed.

Packaging of the DNA into the head does not require a complete length of wild type lambda. It has been determined that a lambda molecule that is between 78% and 105% of wild type length can be packaged. This is from 37 to 53 kb in length. Two important developments suggested that lambda may be suitable as a cloning vector. First it was determined that the gene products between the J and N genes could be removed and the life cycle could be completed. Second, restriction enzyme sites could be eliminated which permitted the development of a vector with a single site for insertion of foreign DNA.

Two types of vectors have been developed:

- *Insertional vector*: DNA is inserted into a specific site.
- *Replacement vector:* Foreign DNA replaces a piece of DNA (stuffer fragment) of the vector.

Let's talk about a specific vector EMBL 3 and EMBL 4. One important concern when cloning with lambda vectors is that you want to maximize the number of resulting phage particles that contain foreign DNA. Or said another way you want to minimize the number of wild type particles. One approach is through spi selection. This refers to sensitivity to P2 interference.

Table. Bacteriophage Phenotypes for Growth on P2 Lysogens

Phenotype	Growth on P2 Lysogens? (bacterial strain)
spi+ (red+gam+)	Poor
spi– (red–gam–)	**Good**

EMBL 3/4 vectors have placed the red and gam genes in the stuffer fragment. Thus only those particles from which the stuffer has been replaced can grow well in a P2 lysogen bacterial cell.

Cloning in Lambda Vectors

Phages are viruses that can infect bacteria. The major advantage of the phage vector is its high transformation efficiency, about 1000 times more efficient than the plasmid vector.

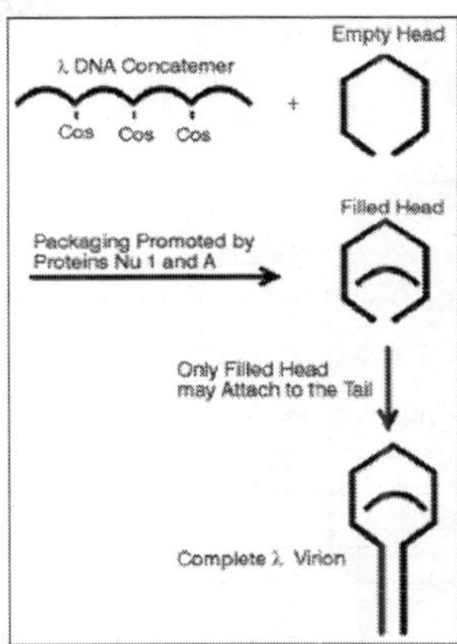

Fig. The Assembly Process of the Virion.

The DNA to be cloned is first inserted into the DNA, replacing a nonessential region. Then, by an in vitro assembly system, the virion carrying the recombinant DNA can be formed. The genome is 49 kb in length which can carry up to 25 kb foreign DNA.The extreme ends of the DNA are known as COS sites, each is single stranded, 12 nucleotides long. Because their sequences are complementary to each other, one end of DNA may base-pair with the other end of a different DNA, forming concatemers. The two ends of a DNA may also bind together, forming a circular DNA.

In the host cell, the DNA circularizes because ligase may seal the join of the COS sites. In the assembly process of virions, two proteins Nu1 and A can recognize the COS site, directing the insertion of the DNA between them into an empty head. The filled head is then attached to the tail, forming a complete virion. The whole process normally takes place in the host cell. However, to prepare the virion carrying recombinant DNA, the following in vitro assembly system is commonly used. Proteins Nu1 and A are encoded by the genes in the genome.

If the two genes are mutated, DNA cannot be packaged into the pre-assembled head. Because tails attach only to filled heads, the cell will accumulate separate empty heads and tails, which can then be extracted. When the extract is mixed with recombinant DNA and proteins Nu1 and A, the complete virion carrying recombinant DNA will be assembled.

COSMIDS

Cosmids are plasmid vectors that contain cos sites. The cos site is the only requirement for DNA to be packaged into a phage particle. Cosmids were developed in light of this observation. How do you clone into cosmid vectors?

- Clone the DNA into the vector as you would with any plasmid.
- Introduce the DNA into the bacterial cell via a phage particle.
- Propagate as plasmid.

Since phage particles can accept between 38 and 53 kb of DNA and since most cosmids are about 5 kb, between 33 and 48 kb of DNA can cloned in these vectors.

It has the following advantages:

- High transformation efficiency.
- The cosmid vector can carry up to 45 kb whereas plasmid and phage vectors are limited to 25 kb.

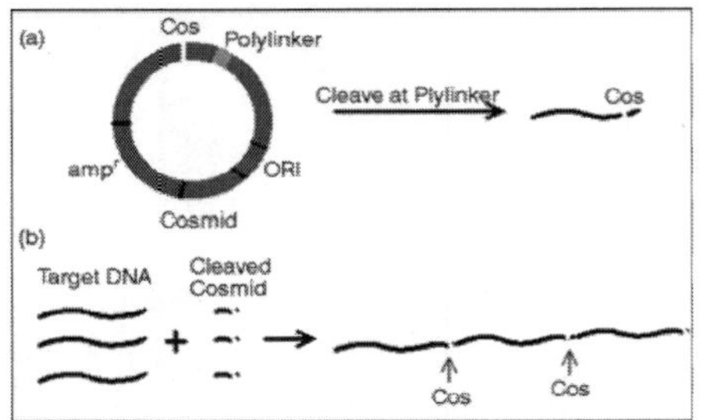

Fig. Cloning by using Cosmid Vectors.

(a) In Addition to Ampr, ORI, and Polylinker as in the Plasmid Vector, the Cosmid Vector also Contains a COS Site. (b) After Cosmid Vectors Are Cleaved with Restriction Enzyme, they are Ligated with DNA Fragments. The Subsequent Assembly and Transformation Steps are the same as Cloning with ? Phages.

Problems Associated with Lambda and Cosmid Cloning

- Since repeats occur in eukaryotic DNA rearrangements can occur via recombination of the repeats present on the DNA inserted into lambda or cosmid.
- Cosmids are difficult to maintain in a bacterial cell because they are somewhat unstable.

YEAST ARTIFICIAL CHROMOSOMES (YACS)

A yeast artificial chromosome (YAC) is a vector used to clone DNA fragments larger than 100 kb and up to 3000 kb. YACs are useful for the physical mapping of complex genomes and for the cloning of large genes. First described in 1983 by Murray and Szostak, a YAC is an artificially constructed chromosome that contains a centromere, telomeres and an autonomous replicating sequence (ARS) element, which are required for replication and preservation of YAC in yeast cells.

ARS elements are thought to act as replication origins. A YAC is built using an initial circular plasmid, which is typically broken into two linear molecules using restriction enzymes; DNA ligase is then used to ligate a sequence or gene of interest between the two linear molecules, forming a single large linear piece of DNA.

A plasmid-derived origin of replication (ori) and an antibiotic resistance gene allow the YAC vector to be amplified and selected for in E.coli. TRP1 and URA3 genes are included in the YAC vector to provide a selection system for identifying transformed yeast cells that include YAC by complementing recessive alleles trp1 and ura3 in yeast host cell. YAC vector cloning site for foreign DNA is located within the SUP4 gene. This gene compensates for a mutation in the yeast host cell that causes the accumulation of red pigment. The host cells are normally red, and those transformed with YAC only, will form colourless colonies. Cloning of a foreign DNA fragment into the YAC causes insertional inactivation, restoring the red colour. Therefore the colonies that contain the foreign DNA fragment are red.

One goal of molecular genetics is to obtain physical data about the genomic organization of long stretches of DNA. Traditionally, this data has been obtained by a technique called chromosome walking. Walking is performed by subcloning the end of a lamda or cosmid clone and screening a library for other clones that contain similar sequence information. If this new clone overlaps a portion of the original clone, then the length of the DNA of interest is extended by the length of DNA in the second clone that is not found in the original clone.

By performing these steps successive times, a long distance map can be obtained. This technique though has difficulties. First, each step is technically slow. Second, if you use lambda or cosmid clones, you might only extend the region of interest by 5-10 kb in each step of the walk. Finally, if any of the clones that are obtained contain repeated sequences, the subclone could lead you to another region of the genome that is not contiguous with the region of interest.

Yeast artificial chromosomes can alleviate some of these problems because of the large (100-1000kb) amount of DNA that can be cloned. First of all, YACs cannot speed up each step of the walk because the subcloning and screening steps can only be performed so quickly. But they can solve the other two problems. Because they carry large amounts of DNA, each step can easily extend the region of interest by 50-100 kb and up to as much as 500 kb. Thus a long distance map of the region can be obtained in several steps. Secondly, although repetitive regions may be 10-20 kb in length they are rarely, longer than 50 kb. Thus a YAC with 100kb will contain some region that is single copy which can be used for further steps in the walk.

Features of YACs:

- Large DNA (>100 kb) is ligated between two arms. Each arm ends with a yeasttelomere so that the product can be stabilized in the yeast cell. Interestingly, larger YACs are more stable than shorter ones, which favours cloning of large stretches of DNA.
- One arm contains an autonomous replication sequence (ARS), a centromere (CEN) and aselectable marker (trp1). The other arm contains a second selectable marker (ura3).

- Insertion of DNA into the cloning site inactivates a mutant expressed in the vector DNA and red yeast colonies appear.
- Transformants are identified as those red colonies which grow in a yeast cell that is mutant for trp1 and ura3. This ensures that the cell has received an artificial chromosome with both telomeres (because of complementation of the two mutants) and the artificial chromosome contains insert DNA (because the cell is red).

DNA cloning is a technique that allows the wholesale production of a specific DNA sequence. DNA containing a gene of interest is inserted into the purified DNA genome of a self replicating element, which can be a plasmid, a virus or in this case, a yeast artificial chromosome (YAC). A YAC can be considered as a functional artificial chromosome (self replicating element), since it includes three specific DNA sequences that enable it to propagate from one cell to its offspring:

- *TEL*: The telomere which is located at each chromosome end, protects the linear DNA from degradation by nucleases.
- *CEN*: The centromere which is the attachment site for mitotic spindle fibers, "pulls" one copy of each duplicated chromo-some into each new daughter cell.
- *ORI*: Replication origin sequences which are specific DNA sequences that allow the DNA replication machinery to assemble on the DNA and move at the replication forks.

It also contains few other specific sequences like:

- *A and B*: selectable markers that allow the easy isolation of yeast cells that have taken up the artificial chromosome.
- Recognition site for the two restriction enzymes EcoRI and BamHI.

While DNA cloning into a plasmid allows the insertion of DNA fragment of about 10,000 nucleotide base pairs, DNA cloning into a YAC allows the insertion of DNA fragments up to 1,000,000 nucleotide base pairs. Why is it so important to be able to clone such large sequences? To map the entire human genome (3x1,000,000,000 nucleotide base pairs) it would require more than 100,000 plasmid clones. In principle, the human genome could be represented in about 10,000 YAC clones.

BACTERIAL ARTIFICIAL CHROMOSOMES (BACS)

A bacterial artificial chromosome (BAC) is a DNA construct, based on a functional fertility plasmid (or F-plasmid), used for transforming and cloning in bacteria, usually E. coli. F-plasmids play a crucial role because they contain partition genes that promote the even distribution of plasmids after bacterial cell division. The bacterial artificial chromosome's usual insert size is 150-350 kbp, but can be greater than 700 kbp. A similar cloning vector called a PAC has also been produced from the bacterial P1-plasmid.Bacterial Artificial

Chromosomes (BAC) have been developed to hold much larger pieces of DNA than a plasmid can. BAC vectors were originally created from part of an unusual plasmid present in some bacteria called the F' plasmid. The F' plasmid allows bacteria to have "sex" (well, sort of: F' helps bacteria give its genome to another bacteria but this only happens rarely when bacteria are under a lot of stress). F' had been studied extensively and it was found that it could hold up to a million basepairs of DNA from another bacteria. Also, F' has origins of replication and bacteria have a way to control how F' is copied. In 1992, Hiroaki Shizuya took the parts of F' that were important, cleaned it up, and turned it into a vector.

BAC vectors are able to hold up to 350 kb of DNA and have all of the tools that a vector needs to work properly, like replication origins, antibiotic resistance genes, and convenient places where clone DNA can insert itself. With these vectors it is possible to study larger genes, several genes at once, or entire viral genomes. By using a vector that can hold larger pieces of DNA, the number of clones required to cover the human genome six times theoretically could drop from 1.8 billion to about 50 million. Researchers have modified BAC vectors to become more convenient to use and more useful in specialized situations. In addition to the antibiotic resistance gene that was added to identify transfected bacteria, a gene was added that enabled the bacteria to turn the colourless substance X-gal/IPTG blue. This substance is found in the chicken-soup-like media on which the bacteria colonizes.

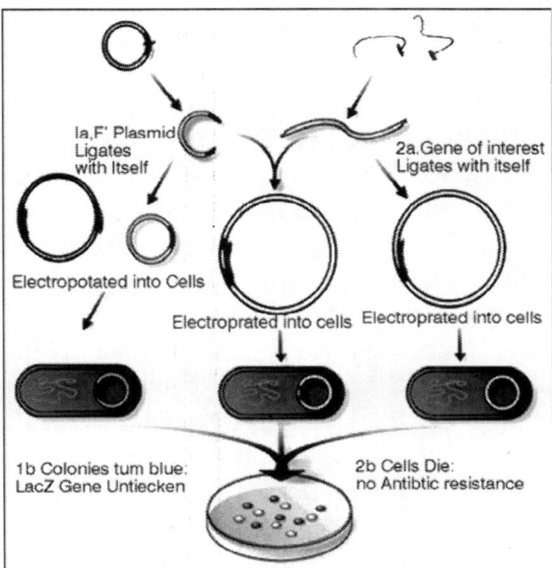

Fig. Selecting for Transformed Bacteria

This colour-changing gene, called lacZ, is split apart when the clone DNA is incorporated into the vector, so it is possible to tell not only if a bacteria had been transfected (meaning incorporated into the cell), but also if the bacteria was transfected with the vector containing insert DNA or just the vector alone (remembering that if the vector has properly incorporated the clone DNA, it will have lost its ability to change X-gal/IPTG).

Another similar modification uses the gene called sacB, which encodes a protein called levansucrase. This protein turns sucrose - table sugar - into levan, a toxic substance to bacteria. In a similar fashion, bacteria grown on media with sucrose will die if sacB is not broken up by inserted DNA. If the vector carries a DNA insert, then sacB is broken and it won't produce levansucrase, so the bacteria can survive in the sucrose media. In theory, only bacteria transfected with a vector containing insert DNA would be able to grow and form colonies. Some modifications to BAC vectors make them more specialized. For instance, there are researchers studying the herpes virus who have made a BAC vector that can be cultured in bacterial cells and then when put into a mammalian cell, instantly releases its insert DNA - in this case, the whole herpes virus genome. With such a vector, it is easier to grow sufficient amounts of the herpes virus for research, since it can live in bacterial cultures, instead of requiring their endemic mammalian cell cultures, which are extremely difficult to maintain. This in turn makes it easier to make modifications to the virus and study what each of its genes do. The creation of BAC vectors has allowed researchers to do many things that they could not do before and do them more quickly and more easily. Some of the scientific practices that have simplified with BAC vectors includes phylogenetic studies (studies which examine species' relationship to one another) and what the absolute minimum size of a genome could be.

The world is currently overrun by a plethora of microbial species that cannot be grown in cultures. BACs have allowed researchers to look at microbial DNA without having to actually grow the organisms, since the DNA is kept within easy-to-grow bacterial cultures. BAC vectors are also useful for studying pathogens, and are helpful in the development of vaccines. Many pathogens are becoming resistant to all antibiotics available to medicine and BAC vectors are playing a tremendous role in discovering new and powerful antibiotics in the environment. Discovering enzymes that are able to help clean up oil spills or help farmers breed healthier farm animals or even process radioactive waste are just a few examples of what Bacterial Artificial Chromosomes can do. BAC vectors are useful tools and the methods for working with them are fairly well developed. Continuing advances such the modification of experimental animals will only increase the wide variety of uses for big bad BACs.

LIBRARY SCREENING AND GENE SEQUENCING

Once a library is constructed it is screened for a particular gene of interest. Screening is based on homology between a probe and one of the clones in the library. The probe is normally a nucleic acid that has some sequence homology to the gene that is represented in the library. The library is the collection of clones from the source DNA which is inserted into a cloning vector. For example, a bean genomic lambda library contains pieces of the entire complement of bean DNA, from 9-20 kb in size, cloned into a lambda vector. Any probe used to screen the library should have some homology to a clone in the bean library, and this homology would allow you to select the appropriate clone from the many clones that do not contain your DNA of interest. Probes come in several forms. The more homologous (similar in DNA sequence) the probe is to the sequence that is being sought, the easier it is to select a clone from the library. For example, a bean lambda library is considered a genomic library because it contains all the DNA sequences found in the bean genome. A bean leaf cDNA library, though, would contain only those sequences that are expressed in the leaf after they have undergone processing.

Thus the cDNA clones would not contain any of the intron sequences or the controlling elements of the gene. To obtain a genomic clone from the bean lambda library, the best probe would be a cDNA clone obtained from screening a bean cDNA library. This type of probe, a probe that contains the exact sequence of the sequence that is being sought, is called a homologous probe. But how was the original cDNA clone obtained so that it could be used as a probe.

The cDNA library from which the clone was obtained might have been screened by a probe from another species which represents coding information for the same gene but from another species. For example, a bean leaf cDNA library may be screened with a tomato RUBISCO small subunit clone. A probe that contains DNA which encodes for the gene of interest but from another species is called a heterologous probe. Many plant genes have been cloned by screening libraries with both homologous and heterologous probes.

Polymerase chain reaction techniques are now commonly used to clone genes. To use this technique primers must be designed that are complementary to your target sequence. One oligonucleotide will be complementary to the anticoding strand (the DNA strand of the gene complementary to the mRNA and used as a template for transcription). The second oligonucleotide will be complementary to the coding strand (the DNA strand of the gene complementary to the anticoding strand).

If the gene has been cloned in another species, you can use that sequence information to design primers to amplifiy a fragment of the gene from the DNA of your species. Often, the gene you have an interest in has never been cloned, but you may have isolated the protein. If this is the case, the first step of this method is to obtain partial protein sequence information of the protein. Micropeptide sequencers are available that can rapidly generate sequence information. From this sequence, you can use reverse translation of the amino acid sequence to obtain the nucleic acid sequences of an amino-terminal fragment and a carboxy-terminal fragment. In this case the fragment complementary to anticoding strand will be a direct conversion of the genetic code for the amino acids in the amino-terminal fragment of the protein. The strand complementary to the coding strand will be complementary to the derived nucleic acid sequence of the carboxy-terminal fragment. Let's say that the following is the N-terminal sequence of the peptide for which you want to derive a synthetic oligonucleotide:

NH_3+–Met–Cys–Val–Lys–Thr–Pro–CO_2–

The following would be an appropriate probe for the gene.

5'-A T G T G T/G G T N A A A/G A G N C C – 3'

(N = A, T, G, and C)

When you are constructing these probes several concepts should be kept in mind. First, the genetic code is written in the mRNA sequence. This synthetic oligonucleotide will thus, be complementary to the anit-coding strand that is used as the template for the mRNA. Secondly, because you only know the amino acid and not the DNA sequence you must deal with redundancies of the genetic code.

For example leucine can be represented by CTA, CTT, CTG, CTC. Thus your synthetic oligonucleotide needs to contain all the possibilities. Therefore the above sequence is actually a mixture of 64 (4×4×2×2) different sequences. Because of this redundancy it is always best to make you oligonucleotide one short of full length. This will help some of the redundancy problems. In this example, the probe is 17 nucleotides long and the redundancies for proline are eliminated. Another method of reducing the redundancy problem is to incorporate the nucleotide deoxyinosine at any position where a high level or redundancy occurs. Deoxyinosine is a modified nucleotide that does not pair with any of the four bases found in DNA. Therefore, it does not affect the homology of the probe.

The following may be the amino acid sequence of a carboxy-terminal fragment:

NH_3+–Phe–Tyr–His–Thr–Val–Ala–CO_2–

The appropriate oligonucleotide sequence from the above sequence for PCR amplification would be:

5'- G C N A C N G T A/G T G A/G T A A/G A A -3'

Amplification of DNA between these two oligonucleotides will provide a homologous probe because it will represent the exact sequence of the gene for which you are searching. Once this probe is radiolabelled by nick-translation (as all probes must be radilabelled before library screening), it can be used to screen a cDNA or genomic library for a clone complementary to the amplified fragment.

DNA Sequencing With the Chain-Termination, Dideoxy Technique of Sanger

Once you have a candidate clone, you will want to sequence it to determine if it is similar to another gene already sequenced or whether it is unique. DNA sequencing is a DNA replication based reaction. The two requirements for DNA replication are a DNA template and a free 3'-OH group. These requirements must be met for any sequencing procedure. The following steps illustrate the Sanger procedure, the most widely used DNA sequencing procedure.

- Although this technique originally used DNA cloned in a vector that can generate single- stranded DNA, sequencing is now routinely performed in double-stranded plasmid or cosmid vectors. pUC plasmids and related plasmids are popular for double stranded sequencing. To begin the procedure, the DNA is made single stranded by a combination of heating and alkaline conditions.
- A primer which has a free 3'-OH group is annealed to the vector. A convenient primer is the multiple cloning site from the pUC plasmids. DNA synthesis will initiate here. 3. Four reactions are performed, each containing dATP, dGTP, TTP and 32P-dCTP and one dideoxy nucleotide in low concentration. Dideoxy nucleotides have a hydrogen rather than a - OH group attached to the #3 carbon and cannot be used for further extension of the chain. Thus in this reaction, various synthesis products will end with that nucleotide.
- Add DNA polymerase to the reaction. Nucleotides will be added one at a time but the chain will terminate when a dideoxy nucleotide is inserted. Because the dideoxy nucleotide is limiting a series of products, each ending with the same base, will be obtained.
- Separate the fragments on an acrylamide gel, develop via autoradiography and read the sequence.

2

Plant Cells

A cell is a very basic structure of all living systems, consisting of protoplasm within a containing cell membrane. Only entities such as viruses—literally on the boundary between non–living chemicals and living systems—lack cells or basic cell structure. All plants, including very simple plants called algae, and all animals are made up of cells, and these are organized in various ways to create structure and function in an organism. Biologists recognize two basic types of cells: prokaryotic and eukaryotic. Prokaryotic cells are structurally more simple. They are found only in single–celled and some simple, multicellular organisms (all bacteria and some algae, which all belong to Bacteria and Archaea domains). Eukaryot cells are found in most algae, all higher plants, fungi, and animals (Eukarya domain). Thus, differences between these two cell types are critical to how an organism is classified, and an important consideration in the evolutionary sequence of life on the planet Earth. Improvements in microscopy, especially development of the Electron microscope, have revealed that cells are not merely membranous sacks containing fluid of gel–like consistency. The degree of organization of the cytoplasm into organelles and their membranes should have you convinced that much (perhaps most) of what is really going on around you on this planet is occurring at a scale that is simply inaccessible to your eyes.

And while you cannot be expected to directly observe chemical reactions at a molecular scale, contemplate that you cannot, even with powerful optics, directly observe most of the structure where these reactions are somehow controlled to produce outcomes favourable to life—indeed, are life. Hopefully, as you acquire knowledge and become a biologist—a botanist—you will learn to recognize the relevant phenomena by their macroscopic expressions (that which you can readily observe with the unaided eye). Plant cells are eukaryotic cells that differ in several key respects from the cells of other eukaryotic organisms.

Their distinctive features include:

- A large central vacuole, a water–filled volume enclosed by a membrane known as the tonoplast maintains the cell's turgor, controls movement of molecules between the cytosol and sap, stores useful material and digests waste proteins and organelles.

- A cell wall composed of cellulose and hemicellulose, pectin and in many cases lignin, are secreted by the protoplast on the outside of the cell membrane. This contrasts with the cell walls of fungi (which are made of chitin), and of bacteria, which are made of peptidoglycan.
- Specialised cell–cell communication pathways known as plasmodesmata, pores in the primary cell wall through which the plasmalemma and endoplasmic reticulum of adjacent cells are continuous.
- Plastids, the notables one being the chloroplasts, which contain chlorophyll and the biochemical systems for light harvesting and photosynthesis, but also amyloplasts specialized for starch storage, elaioplasts specialized for fat storage, and chromoplasts specialized for synthesis and storage of pigments. As in mitochondria, which have a genome encoding 37 genes, plastids have their own genomes of about 100–120 unique genes and, it is presumed, arose as prokaryotic endosymbionts living in the cells of an early eukaryotic ancestor of the land plants and algae.
- Unlike animal cells, plant cells are stationary.
- Cell division by construction of a phragmoplast as a template for building a cell plate late in cytokinesis is characteristic of land plants and a few groups of algae, the notable one being the Charophytes and the Order Trentepohliales
- The sperm of bryophytes have flagellae similar to those in animals, but higher plants, (including Gymnosperms and flowering plants) lack the flagellae and centrioles that are present in animal cells.

CELL TYPES

Parenchyma cells are living cells that have diverse functions ranging from storage and support to photosynthesis and phloem loading (transfer cells). Apart from the xylem and phloem in its vascular bundles, leaves are composed mainly of parenchyma cells.

Some parenchyma cells, as in the epidermis, are specialized for light penetration and focusing or regulation of gas exchange, but others are among the least specialized cells in plant tissue, and may remain totipotent, capable of dividing to produce new populations of undifferentiated cells, throughout their lives. Parenchyma cells have thin, permeable primary walls enabling the transport of small molecules between them, and their cytoplasm is responsible for a wide range of biochemical functions such as nectar secretion, or the manufacture of secondary products that discourage herbivory. Parenchyma cells that contain many chloroplasts and are concerned primarily with photosynthesis are called chlorenchyma cells. Others, such as the majority of the parenchyma cells in potato tubers and the seed cotyledons of legumes, have a storage function.

Collenchyma cells—collenchyma cells are alive at maturity and have only a primary wall. These cells mature from meristem derivatives that initially resemble parenchyma, but differences quickly become apparent. Plastids do not develop, and the secretory apparatus (ER and Golgi) proliferates to secrete additional primary wall.

The wall is most commonly thickest at the corners, where three or more cells come in contact, and thinnest where only two cells come in contact, though other arrangements of the wall thickening are possible.

Pectin and hemicellulose are the dominant constituents of collenchyma cell walls of dicotyledon angiosperms, which may contain as little as 20% of cellulose in Petasites. Collenchyma cells are typically quite elongated, and may divide transversely to give a septate appearance. The role of this cell type is to support the plant in axes still growing in length, and to confer flexibility and tensile strength on tissues.

The primary wall lacks lignin that would make it tough and rigid, so this cell type provides what could be called plastic support—support that can hold a young stem or petiole into the air, but in cells that can be stretched as the cells around them elongate. Stretchable support (without elastic snap–back) is a good way to describe what collenchyma does. Parts of the strings in celery are collenchyma.

- *Sclerenchyma Cells*: Sclerenchyma cells (from the Greek skleros, hard) are hard and tough cells with a function in mechanical support. They are of two broad types—sclereids or stone cells and fibres. The cells develop an extensive secondary cell wall that is laid down on the inside of the primary cell wall. The secondary wall is impregnated with lignin, making it hard and impermeable to water. Thus, these cells cannot survive for long' as they cannot exchange sufficient material to maintain active metabolism. Sclerenchyma cells are typically dead at functional maturity, and the cytoplasm is missing, leaving an empty central cavity.

Functions for sclereid cells (hard cells that give leaves or fruits a gritty texture) include discouraging herbivory, by damaging digestive passages in small insect larval stages, and physical protection (a solid tissue of hard sclereid cells form the pit wall in a peach and many other fruits).

Functions of fibres include provision of load–bearing support and tensile strength to the leaves and stems of herbaceous plants. Sclerenchyma fibres are not involved in conduction, either of water and nutrients (as in the xylem) or of carbon compounds (as in the phloem), but it is likely that they may have evolved as modifications of xylem and phloem initials in early land plants.

TISSUE TYPES

The major classes of cells differentiate from undifferentiated meristematic cells (analogous to the stem cells of animals) to form the tissue structures of roots, stems, leaves, flowers, and reproductive structures.

Xylem cells are elongated cells with lignified secondary thickening of the cell walls. Xylem cells are specialised for conduction of water, and first appeared in plants during their transition to land in the Silurian period more than 425 million years ago. The possession of xylem defines the vascular plants or Tracheophytes. Xylem tracheids are pointed, elongated xylem cells, the simplest of which have continuous primary cell walls and lignified secondary wall thickenings in the form of rings, hoops, or reticulate networks.

More complex tracheids with valve–like perforations called bordered pits characterise the gymnosperms. The ferns and other pteridophytes and the gymnosperms have only xylem tracheids, while the angiosperms also have xylem vessels. Vessel members are hollow xylem cells aligned end–to–end, without end walls that are assembled into long continuous tubes. The bryophytes lack true xylem cells, but their sporophytes have a water–conducting tissue known as the hydrome that is composed of elongated cells of simpler construction.

Phloem is a specialised tissue for food conduction in higher plants. The conduction of food is a complex process that is carried in the plant with the help of special cell called phloem cells. These cells conduct inter–and intra-cellular fluid (food—proteins and other essential elements required by the plant for its metabolism) through the process of osmosis. This phenomenon is called ascent of sap in plants.

Phloem consists of two cell types, the sieve tubes and the intimately-associated companion cells. The sieve tube elements lack nuclei and ribosomes, and their metabolism and functions are regulated by the adjacent nucleate companion cells. Sieve tubes are joined end–to–end with perforate end–plates between known as sieve plates, which allow transport of photosynthate between the sieve elements. The companion cells, connected to the sieve tubes via plasmodesmata, are responsible for loading the phloem with sugars. The bryophytes lack phloem, but moss sporophytes have a simpler tissue with analogous function known as the leptome.

Plant epidermal cells are specialised parenchyma cells covering the external surfaces of leaves, stems and roots. The epidermal cells of aerial organs arise from the superficial layer of cells known as the tunica (L1 and L2 layers) that covers the plant shoot apex, whereas the cortex and vascular tissues arise from innermost layer of the shoot apex known as the corpus (L3 layer). The epidermis of roots originates from the layer of cells immediately beneath the root cap. The epidermis of all aerial organs, but not roots, is covered with a cuticle made of waxes and the polyester cutin. Several cell types may be present in the epidermis. Notable among these are the stomatal guard cells, glandular and clothing hairs or trichomes, and the root hairs of primary roots. In the shoot epidermis of most plants, only the guard cells have chloroplasts. The epidermal cells of the primary shoot are thought to be the only plant cells with the biochemical capacity to synthesize cutin.

GROWTH OF PLANT CELL

The plant cell after formation normally increases very considerably in size. This process may be studied in two situations:

- When the cells are in intact organs,
- When they are in isolated systems.

The first indicates the scope and characteristics of the process, the second the factors that control the several aspects of it. The two groups of studies based on the different situations complement each other, and it is therefore proposed to consider below, first, observations made with intact organs, and secondly, observations made with isolated fragments. This treatment of plant cell growth is not intended to be exhaustive and one topic in particular is omitted.

For two reasons no discussion is attempted of the nature of the auxin effect in cell growth. The subject is omitted, since it has recently been discussed in some detail by Audus and, secondly, since it is possible that an elucidation of this problem can only be developed after the nature of cell growth has been more fully explored.

Hitherto plant cell growth has frequently been described as cell extension. In our view this term is an unfortunate one. It is probably a legacy of the period when cell growth was considered a simple matter of an inflation due to the absorption of water.

Recent work, however, has shown that the increase in size of the cell involves increases in many components of the system, and we therefore prefer the term cell growth to the older one of cell extension.

PROCESS OF CELL GROWTH

The process of cell growth involves the changes that occur between the meristematic and the mature fully vacuolated state. The meristematic cell is small isodiametric and without a prominent central vacuole. As the cell enlarges a central vacuole develops and the shape changes. Eventually a large parenchymatous cell is established with a thick wall and a thin peripheral layer of cytoplasm. This set of changes is readily studied on the root. It is probable that this is a more satisfactory experimental object than the coleoptile, since with this organ the earlier stages of growth are not readily accessible to investigation. The root, being an organ of unlimited growth, has an apical meristem from which cells are being continuously generated. These cells immediately they are formed commence enlarging, and thus at increasing distances from the apex the tissue is composed of cells in progressively advanced stages of development. A similar situation is involved in the apex of the stem. Here, however, the position is complicated by the development of lateral members from the surface of the meristematic dome.

In the root the surfaces of the growing regions are not covered with developing primordia, and this circumstance has been exploited in the development of two sets of techniques for the study of cell growth. Since the surface of the growing region is naked in the root direct observations may be made microscopically on epidermal cells.

The dimensions of epidermal cells may be measured at increasing distances from the apex and the growth of these cells determined over the whole range of development. This technique has been used in principle by Burstrom, Brumfield and by Goodwin.

It has been shown that sliding growth is not involved, and changes in surface cells may therefore be taken as representative of corresponding changes occurring in deeper tissues. This general technique has yielded important results, and it has the merit that it involves observations on a more or less uniform tissue.

It carries the limitation, however, that only a limited variety of observations can be made with it. It cannot, for instance, yield data regarding metabolic conditions in developing cells. In order to avoid this difficulty another experimental approach has been proposed by Brown and Broadbent.

These workers cut successive sections from the apex backwards, made observations on each section, and related the values obtained to the total number of cells in each section. They used a technique for determining the number of cells in a tissue originally developed by Brown and Rickless.

The range of observations that can be made with this technique is clearly greater than that possible with the earlier procedure. It involves, however, two possible sources of error. Cell division does not cease at the same distance from the apex in all regions of the root, and all cells in the section are therefore not exactly at the same stage of development. Secondly, sectioning necessarily involves a destruction of cells the extent of which cannot easily be estimated. The first source of error is not decisive, since the distance over which cell division ceases is small relative to the total distance over which growth occurs, and the average given by the section therefore represents approximately the position for all the cells in the section. The error from the destruction of cells can be reduced by increasing the length of the section and thus decreasing the proportion of damaged cells. These data are from the observations of Brown and Broadbent. The full development occurs over a distance of about 5.0 mm. from the apex and involves about a twenty- to thirtyfold increase in volume. The enlargement is accompanied by important metabolic changes. In the present connexion it may be noted that the enlargement in the early stages is accompanied by the gradual suppression of the capacity to divide.

The zone that stretches from the tip to about 1.5 mm. behind it is that over which mitotic figures are normally distributed. At the same time the observations of Wagner and of Gray and Scholes have shown that the mitotic frequency is greatest at the tip and decreases progressively with increasing distance from it.

Brown has emphasized that the frequency of mitotic figures may not be an index of the relative rate of division. In this case, however, since the frequency decreases to zero it probably is, and the data therefore suggest that after formation at the apex the increase in the average cell size is accompanied by a corresponding increase in the average length of the interphase. As growth proceeds the probability of division decreases, and at a certain point the probability becomes so small as to constitute virtual cessation of the process.

The conditions that determine this transition have not been explored, but it is clear that the fact that it occurs indicates that growth in the early stages is accompanied by metabolic changes that tend to suppress division. At about the point at which division ceases a prominent central vacuole appears in the cell. This vacuole, which is probably a development from a dispersed vacuolar system in the meristematic cell, increases in size with the enlargement of the cell.

During growth, however, the percentage water content is increasing, indicating that the volume of the vacuole is increasing more rapidly than the other components, and that the enlargement of the cell is being determined primarily by the absorption of water. At the same time it is evident that the increase in water content is being sustained by relatively smaller increases in other phases of the system. These data indicate that the dry weight of the cell increases about tenfold during the course of growth.

Preston and Clark and Wirth have shown that the cellulose content of cells in the coleoptile increases during growth and Wirth has demonstrated corresponding changes with soluble carbohydrates. The increase in dry weight of the cell in the root is therefore probably due partly to similar changes. It is evident, however, that it is not due only to changes in these constituents. In 1941 Blank and Frey-Wyssling described a series of important observations on cell growth in the coleoptile of maize. They reported an increase in protein content during the enlargement of the cell. Blank and Frey-Wyssling found similar changes occurred during cell enlargement in the hypanthium of *Oenothera*. Kopp calculated that in the root the protein content increased, and Brown and Broadbent demonstrated this condition experimentally in pea roots. The appropriate curve is reproduced from the data of Brown and Broadbent, and it is evident from this that protein increases progressively during growth of the cell, resulting in at least a fivefold increase. At the same time it may be noted that when growth ceases there is a slight but significant decrease in protein, the final level being lower than it is when growth stops, although greater than it is in the meristematic region. The increase in protein necessarily implies an increase in the general metabolic activity of the cell during growth. Kopp calculated and Brown and Broadbent later demonstrated an increase in respiration rate per cell during growth.

The data of Brown and Broadbent suggest, however, that when growth ceases there is a slight fall in respiration which is coincident with the decrease in protein. The increase in protein content might also be expected to increase the quantities of particular enzymes. We have recently been examining the changes in the activities of certain enzymes during the growth of the cell.

These data have been obtained by taking successive sections along bean roots, killing the sections by freezing at - 10°C., determining on the killed tissue the activities of certain enzymes, and finally reducing the values for each section to a unit cell basis. The activities of three enzyme systems have been investigated: a dipeptidase which hydrolyses alanylglycine, an invertase, and an acid phosphatase.

The activity of the dipeptidase at increasing distances from the apex of the root of barley has already been examined by Bottelier, Holter and Linderstrom-Lang, and Wanner and Leupold have determined the relative sucrose-splitting capacity in different regions of the root of maize. In the bean roots growing in the conditions of our experiments growth continues over about the first 8 mm. The data show that over this range there is about a fivefold increase in dipeptidase, about a twofold increase in phosphatase, and about a twentyfold increase in invertase activity.

Further, it is evident that in about the region where growth in length is ceasing the activities of the respective enzyme systems decrease. Determinations of protein in cells of bean roots are not available, but in view of the earlier observations with peas it is probable that the increase in enzyme activity is due to the differentiation of part of the protein increment into enzyme bodies, and that the decrease in enzyme is similarly partly due to a decrease in protein. It is an important aspect of the situation that the activities of the three enzymes do not change relatively to the same extent. The relative increases are different and the ratios of the activities of any two of the enzymes therefore change during the course of growth. A similar position is indicated by some observations of Linderstrom-Lang and Holter on the activities of what may be two distinct enzymes cleaving leucyl-glycine and alanylglycine respectively at increasing distances from the tip of the root of barley.

The data, although they are not expressed in terms of unit cell number, nevertheless show that the ratio between the two enzymes changes as the distance increases. Clearly since the ratios between the enzyme activities change it is evident that the metabolic pattern must change during the course of growth. During the growth of the cell a differentiation probably occurs in the metabolic state such that the relative intensities of different reactions change in the course of the process. In particular, these observations on the enzyme complement suggest that the metabolic activities of the young cell are different from those of the mature. This conclusion is consistent with the results of an investigation by Morgan of the amino-acid composition of the proteins in different parts of the growing zone of the bean root.

The proteins from sections taken at increasing distances from the apex of the root have been hydrolysed and the hydrolysates examined chromatographically. It is evident that the composition of the proteins varies considerably. Such differences in the composition of the proteins suggest of course corresponding differences in the generalized metabolic pattern. This general conclusion is also consistent with a variety of observations by other workers. Berry and Brock have reported that the respiration of the tip of the onion root is more sensitive to cyanide than is that of the mature tissue, and Kopp has found that per unit protein it is lower at the tip than it is in the mature regions. The data of Brown and Broadbent suggest that the increase in protein per unit increase in volume is greater in the earlier than it is in the later stages of growth.

Finally, Burstrom has shown that the reactions of root cells to sugar, temperature and heteroauxin are different in the early and late stages. The growth of plant cells is interesting on account of the fact that the protoplasm is covered with a solid wall. Any irreversible increase of the cell size involves the growth in area of this wall, which consists partly of crystallized substances. The development of cell walls, especially in the cells of higher plants, passes through two different stages. As long as there is growth in area, they remain thin and flexible. This fine membrane is called the *primary cell wall*, and it is said to grow in area by *intussusception*. As soon as the final shape and size of the cell are reached, the membrane is strengthened by adding new layers to the primary wall. During this second stage the cell wall grows in thickness by *apposition*. The combined layers are called the *secondary cell wall*. The thickness of this is often so large that the slender primary wall is overlooked or neglected, especially in materials of technical importance such as wood pulp, textile fibres, etc. Indirect methods of investigation, such as polarized light and X-rays, indicate that the primary wall consists of a loose submicroscopic network of crystallized strands of cellulose or chitin, arranged in a dispersed texture.

In the secondary wall the framework is much more compact. This compactness is obtained by a parallel arrangement of the submicroscopic strands forming a highly anisotropic *parallel texture*. Two different types of growth in the area of the primary wall can be distinguished:

TIP GROWTH

Tip growth of filamentous cells such as fungal hyphae, hairs (*e.g.* root hairs), latex tubes, pollen tubes, etc. In this case, only the tip of the cell is involved in growth, and no intussusception occurs away from the end. Sometimes the tip growth of very long cells (e.g bast fibres of ramie, which reach a length of over 20 cm.) continue after the formation of the secondary wall has already started in the middle body of the cell.

EXTENSION GROWTH OF CELLS IN ELONGATING TISSUES

Extension growth of the cells in elongating tissues. In this case the whole length of the primary wall is said to increase in area by intussusception. There is an enormous bibliography on this subject, because the extension growth is stimulated by auxin, which was considered at the time as the hormone of the extension growth.

However, the question how auxins intervene in this process cannot be solved until the mechanism of the growth in area has been cleared up. For a long time it was thought that the extension growth consisted of simple water intake, and the turgor stretch of the cell wall caused thereby.

But it has been shown that this type of cell wall growth is much more complicated, since it involves respiration, biosynthesis, and morphogenesis. Moreover the work needed for the elastic deformation of the growing cell wall is negligible as compared with the total energy available for growth. As extension growth concerns tissues with numerous cells which expand simultaneously at different rates, the geometrical problem how contiguous polyhedra can change their shape without being individualized arises. Krabbe thought that a disintegration of the tissue was inevitable and suggested the hypothesis of *sliding growth*.

This implies that the cells of a tissue assume the individuality of independent unities growing along each other like unicellular protophytes. This is an extreme interpretation of sliding growth it is true, but it demonstrates clearly that such an idea is theoretically unsound. Priestley published other arguments against the possibility of this type of growth; sliding of the two adjacent cell walls would interrupt the channels of pits, so that their orifices would no longer correspond.

In consequence one might expect to find a large number of unilateral pits in tissues in which sliding growth had occurred. As this is not the case, Priestley postulated that the adjacent cell faces must grow together at the same rate, and he termed this *symplastic growth*. His theory cannot, however, explain how latex tubes and other specialized cells grow across tissues and thereby come in touch with cells which previously were not their neighbours. This frequently occurs when derivatives of the cambial cells, such as tracheids and fibres, elongate in the tissue. Two adjacent cells are pushed apart by the tip of a cell which intrudes between them. This has been termed intrusive growth by Sinnot and Bloch or interposition growth by Schoch-Bodmer. The last author emphasizes that only the tips of elongating fibres behave in this way. According to her, the walls of the central body of the cell do not grow any longer, so that the entity of the tissue is secured. Pits are only found in that part of the cell wall, and the fibre tips are free from them.

From this description it follows that different parts of the cell wall must behave differently. A plant cell does not grow as a whole, but in a differentiating cell there are growing and full-grown parts at the same time.

STRUCTURE OF CELL WALL

The basic unit in the structural organization of the cellulosic framework of the plant cell wall is the cellulose molecule. Cellulose molecules are not units of definite molecular weight but long chains of varying length formed by the condensation of at least a 100-120 and possibly many times this number of /3 d-glucose molecules1. Studies of cellulose walls with the X-ray and polarized light have furnished evidence that the molecules of cellulose are aggregated into bundles known as micelles.

The micelles have been estimated upon the basis of X-ray photographs to have a diameter of approximately 5 or 6 mp and a length of about 6o m. A unit of this size would consist of about sixty parallel cellulose chains each being made up of about 120 glucose units. It is probable that in many cell walls long chain-like molecules formed by the condensation of other sugars are associated with the cellulose chains in the micelles. Originally the micelles were believed to be well defined units which were cemented together by some non-crystalline material. Recent work indicates however, that many of the cellulose chains are much longer than the micellar aggregates and extend from one micelle to another, thus welding the micelles into a coherent an astomosing system.

It is also probable that the micellar aggregates vary in size. Cellulose walls are no longer regarded, therefore as a system of discrete crystalline units bound together by some cementing substance but rather as a structure composed of molecular aggregates are welded together by interlocking cellulose molecules.

There seems to be little doubt regarding the presence of intermicellar spaces. The spaces form an interconnecting system between the anastomosing micelles.

Black areas represent intermicellar spaces; white areas represent the cellulose micelles.

(A) Cross section,

(B) Longitudinal section.

In walls that are almost pure cellulose, such as the secondary wall of cotton fibres, it is possible that the intermic-ellar spaces are filled chiefly with water.The smallest visible units of cellulose walls are delicate thread like strands or fibrils. In primary walls these fibrils form a loose anastomosing network the meshes of which are usually filled with colloidal pectic compounds.

In secondary walls the fibrils are often grouped into coarser strands which wind around the cell in a steep spiral the angle of which may vary indifferent layers and even in different parts of the same layer.

PHYSICAL PROPERTIES OF CELL WALLS

The physical properties of primary cell walls differ from those of secondary walls. These differences may be ascribed to the greater abundance of cellulose in secondary walls and to differences in the structural organization of the cellulose in the two kinds of walls. Both primary and secondary walls are transparent to wavelengths of the visible spectrum and both are usually quite pcrmeable to most substances dissolved in water that compare favourably with those of steel.

The breaking strength of a flax fibre, for example, may be as great as 110 kg. per mm.2 of wall area while the tensile strength of a hardened spring steel ranges between 150 and I 70 kg. per mm2. Although the tensile strength of primary walls is considerably less than that of secondary walls it is only rarely that they are subjected to strains in excess of their breaking strength. Primary cell walls are usually quite elastic.

The mesophyll cells of the leaves of some species, for example, are known to undergo reversible changes in volume of 30 per cent or more in response A to changes in turgor pressure. Secondary walls, on the other hand, are elastic only within very narrow limits and break suddenly when their tensile strength is exceeded. The elasticity of primary walls is undoubtedly related to the loose open mesh work of cellulose strands that compose it, while the close parallel orientation of the cellulose micelles in secondary walls prevents any appreciable elongation without rupture.

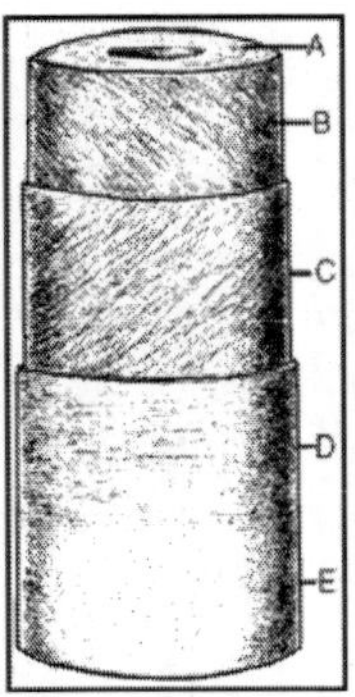

Fig. Diagram Illustrating the Structure of a Thickened Plant Cell Wall.

CHEMICAL CONSTITUENTS OF CELL WALLS

Cellulose is by far the most abundant compound found in the cell walls of the higher plants. Associated with cellulose in all the primary cell walls of vascular plants are greater or lesser amounts of pectic compounds. Their presence in the middle lamella and in the intermicellar spaces of the primary walls has already been noted. Lignin is an important constituent of the walls of most of the cells that make up woody tissues and it also occurs commonly in other thickened walls.

Although lignin can be isolated from cell walls by suitable treatments as a brownish amorphous substance it is probably altered considerably in the process. Its structural formula and molecular weight are unknown. Since the lignins produced in different species appear to differ in their chemical properties it seems very probable that lignin is not a compound of definite molecular weight but a complex mixture of a number of chemically similar substances. The lignin of spruce wood has been extensively studied and several workers have suggested that its molecular weight lies between 800 and 900.

Lignin first appears in the middle lamella and primary wall and later canbe detected in the secondary wall. When associated with cellulose, lignin is present in the spaces between the micelles and not within the micelles. Lignifled walls are usually freely permeable to water and solutes. The tensile strength of lignified cell walls is the same as that of cellulose walls but lignified walls resist compression better than cellulose walls. The increased resistance of lignified walls to compression is explained by the assumption that the presence of lignin in the intermicellar spaces welds the cellulose micelles into a single coherent mass and thus prevents the bending and buckling of the cellulose strands when they are subjected to compression strains.

Cutin is the name applied to the mixture of wax-like materials found on the outer surface of the epidermal cell walls of leaves, stems, fruits, and other organs. These wax-like substances are intimately associated with cellulose and often with pectic substances producing a wall of great structural complexity. Cutinized cell walls are relatively impermeable to water. The presence of cutin in the outer walls of epidermal cells greatly reduces the evapouration of water from the surfaces of plant tissues. Subcrin is similar in many of its properties to cutin.

It constitutes an important part of cork cell walls and it is also found in the walls of a few other specialized types of cells. Most of the surface of perennial plants, aside from the leaves and very young stems is covered with suberized cell walls. Such walls are relatively impermeable to water. Hemicell-uloses are a poorly defined group of poly saccharides associated with cellulose in plant cell walls. They are not chemically related to cellulose, as the name implies, but possess very different chemical and physical properties. Callose is the name given to a carbohydrate membrane substance found in the perforated septa (sieve plates) of the sieve tubes. Similar material presumably of the same chemical composition, has been found in pollen grains and constitutes the inner layer of pollen tubes. It has also been reported as occurring in the fungi. The exact chemical composition of callose is unknown since it has never been obtained in sufficient amounts to permit quantitative determinations. Chitin, a nitrogenous substance common in the exoskeleton of insects, is a constituent of the walls of many fungi and bacteria. It has been reported as being present in the walls of certain algae but it is unknown in any of the higher plants. Tannins are commonly found in the cell sap but they also occur in the walls of certain tissues, especially cork and wood cells.

Mucilages are common constituents of the outer walls of many water plants and occur also in the outer walls of some seed coats, in glandular hairs and in other specialized tissues. Inorganic compounds such as silica and salts of calcium, iron, and other metals are also present in some plant cell walls. None of these inorganic compounds, however, are regarded as essential constituents of the cell wall.

PROTOPLASM

In most mature cells of the vascular plants protoplasmis present only as a thin layer covering the inner surface of the cell walls but in some specialized cells branching strands of protoplasm also extend across the vacuole. Under high magnification the protoplasm of active cells appears as a colourless fluid, in which are suspended numerous tiny granules and droplets of insoluble materials. These granules frequently exhibit active Brownian movement. The fluid component of protoplasm also is frequently in motion, streaming around the inner surfaces of the cell walls. Embedded in this fluid component, and carried passively by it, are the specialized bodies known as plastids.Although protoplasm appears to be a simple liquid, no simple liquid could possibly possess the remarkable powers of synthesis, assimilation, reproduction growth and sensitivity that characterize the protoplasm of living plant cells.

The properties and behaviour of protoplasm clearly show that it is not a substance but that it must be regarded as a complex system of substances. This system is dynamic; it is constantly undergoing changes yet at the same time the changes are so regulated and controlled that the system is not disrupted. A cell is alive only so long as the organization of this dynamic protoplasmic system is maintained.

The protoplasmic system of living plant cells almost invariably contains a well differentiated globular body known as the nucleus. All of the protoplasm outside of the nucleus is designated as cytoplasm. Although the nucleus is separated structurally from the cytoplasm by a delicate membrane the two components are not physiologically isolated. The nucleus and the cytoplasm appear to be mutually dependent and there is good reason to believe that the nucleus controls and regulates the physiological processes that occur in the cytoplasm. Probably because of the close relationship between the nucleus and the cytoplasm many biologists use the terms protoplasm and cytoplasm synonymously.

THE CHEMICAL COMPOSITION OF PROTOPLASM

Since protoplasm is a dynamic system of substances it is not possible to subject it to chemical analysis without destroying it. In the strict sense, therefore, it is impossible to discover the chemical composition of protoplasm. It is possible, however, to examine the substances present after the protoplasmic system has been destroyed and to determine their chemical composition and relative abundance. A number of such studies have been made.

Water is the chief component of all physiologically active plant protoplasm usually making up more than 90 per cent of the system. The water content of the protoplasm of dry seeds, on the other hand, may be less than 10 per cent. Most attempts to determine the chemical composition of plant protoplasm have been made upon species of the myxomycetes. At certain stages in their life history these organisms consist of naked masses of labile protoplasm. They are often found flowing over rotten logs in damp woods.

The fact that the myxomycetes provide relatively large quantities of protoplasm entirely free from cell wall material has made them a favourite object for chemical analysis. Even in such organisms, however, not all of the constituents of the plant body can be regarded as integral parts of the protoplasm. Distributed throughout the protopl-asmic mass are particles of foods and other inert materials which cannot be separated from the protoplasm. As shown in this analysis proteins and other nitrogen-containing compounds constitute the bulk of the organic matter in the plasmodium of this species. Many different varieties of proteins are known to occur in the protoplasm of plant cells. They are compounds of enormous molecular weight and undoubtedly make up a large proportion of the labile structural framework of the protoplasm.

Lipids constitute a smaller fraction of the protoplasm than the proteins. Three types of lipids occur in the protoplasm; the true fats (oils), the phosphatides (phospholipids) and the sterols, of which phytosterol is an example. The oils are generally suspended in the protoplasm in the form of minute globules. They are probably more important as food reserves than as actual constituents of the protoplasm. The phospho-lipids and sterols, on the other hand, are believed to be essential constituents of the protoplasmic system.

The water-soluble carbohydrates, amino acids, etc., present in the plasmodium of this species are probably almost entirely foods. The inorganic compounds (mineral matter) in plant cells are chiefly the phosphates, chlorides sulfates, and carbonates of magnesium, potassium, sodium, and calcium. It supplies no more information regarding the organiz-ation of protoplasm, however, than a chemical analysis of the ground up debris of a wrecked house would furnish regarding the structure of that house. This point is worthy of emphasis, since with the progress of biology it becomes clearer and clearer that the properties of protoplasm are as much a function of its physiochemical organization as of the specific kinds of compounds present.

Although we commonly refer to protoplasm as the essential constituent of all living cells, it is evident that there are at least as many different varieties of protoplasm as there are species of plants and animals. All of them, however, are dynamic systems composed of the same types of compounds, and all of them possess a colloidal organization of a complex type.

STRUCTURE OF GROWING CELL WALLS

The electron microscope shows that young cell walls have a woven texture of dispersed cellulose microfibrils. The diameter of the microfibrils involved measures about 250 A. However, the electron microscope is said to be unsuited for research in cell physiology and growth because the specimens must be examined completely dried *in vacuo*.

Nevertheless, it is possible to take advantage of the results obtained by electron microscopy if they are properly combined with those of other methods. The natural state of the cellulosic framework of the living cell wall can be reconstructed from the electron micrograph if its chemical composition is known. In growing coleoptiles we have found 92.5 per cent water. If this water content is taken as an average for all cell constituents, the cell wall contains only 7.5 per cent dry matter. In corn coleoptiles 32 mm. in length twothirds of this dry material is composed of substances other than cellulose, such as pectins, hemicelluloses, wax. Therefore, the growing cell wall consists of only 2.5 per cent cellulose by weight.

As the density of crystallized cellulose is about 1.55 this is even less by volume. It shows how astonishingly loose is this framework in a growing cell wall, including plenty of space for living cytoplasm and all kinds of incrusting substances. Only when this highly hydrated, spacious network is freed from non-cellulosic substances and completely dried does it furnish the picture of a dense interwoven texture in the electron microscope. The impression of a rather compact framework is further enhanced by the high focal depth of the electron microscope, which yields a sharp picture of microfibrils situated in rather distant planes.

Whereas it is difficult to understand how a woven texture as shown by the electron micrograph can grow by intussusception, where new microfibrils can readily be interlaced. Another handicap to electron microscopy of cell walls has been the method of preparation, which previously consisted in disintegrating cells in a blendor whereby pieces thin enough for electron transmission were obtained.

From these specimens it was not possible to locate the pieces examined in the original cell wall, and the micrographs obtained did not show whether the whole wall of a cell or only parts of it are involved in growth. Muhlethaler showed, however, that the purified cell walls of macerated meristems are thin enough for direct examination without cutting them into smaller bits. This is possible because in these membranes he cellulose frame represents only about 2.5 per cent by weight of the living wall, which has a diameter of less than 1¼ in the swollen state.

So the dried cell wall freed of all incrusting substances would be only about 0.025¼ thick, if its mass were spread as a homogeneous film. Yet, according to the porous structure of the wall the dried framework is correspondingly thicker, say 0.075 ¼. This would correspond to a woven texture of about three microfibrils thick. As macerated collapsed cells consist of two superposed walls the thickness of the preparation may amount to 0.15¼, which is an appropriate value for the examination in the electron microscope. The possibility of investigating whole cells proved an opportunity for observing the same spot of the wall alternatively in the electron and in the ordinary microscope.

For this purpose the walls are stained with benzoazurin, and the resulting dichroism is examined in polarized light. As the direction of the strong absorption coincides with the main trend of the microfibrils, a mutual check on the observations in the electron microscope and the results of this indirect method is possible. Thereby an identification of different cell types and the investigation of individual faces of a cell wall are feasible. Our first intention was to elucidate the tip growth in order to observe the formation of new microfibrils at the very tip of the cell where intuesusception growth occurs. *Root hairs* seemed a favourable object for this purpose, because their tip growth is very fast.

This research yielded the discovery that the tip of root hairs (*Zea mays*) is covered with a felt-like layer of cellulosic microfibrils. The felt is more than 1 ¼ thick and corresponds to the slime layer which covers the growth region of the hair. The microfibrils are anchored in the interwoven cell wall. The slimy consistency of this layer is due to a large amount of interfibrillar substances with high swelling power, such as pectins. These can easily be demonstrated by pectin dyes (ruthenium red, methylene blue), whilst microchemical reactions, aiming at proving the presence of cellulose, fail.

This is probably due to the fact that the microfibrils found are so well crystallized that no intermicellar spaces are left for the penetration of cellulose dyes. The tip of the root hair is so woolly on account of this extracellular felt that the formation of new microfibrils cannot be observed. It is probable, however, that the cytoplasm which synthesizes the wall substances soaks this system of cellulose microfibrils, which on account of the hydrated state of the pectins must be very loose. As the process of neoformation of microfibrils is hidden in the root hairs, we looked for objects where the growth zone is not covered by slimes, such as hyphae of fungi and pollen tubes. A favourable object is the growing *sporangiophore of Phycomyces* before the tip is inflated for the formation of the sporangium. The chitinous cell walls of fungi have the same microfibrillar structure as those of higher plants; the only difference is that the microfibrils consist of chitin instead of cellulose. Curiously enough both types of microfibrils have approximately the same diameter and both are arranged in loosely woven dispersed or compact parallel textures.

At the tip of the young sporangiophore the texture of the microfibrils is very dense, and is somewhat obscured by some incrusting substance which is difficult to remove. The same difficulty is encountered with the tip of *pollen tubes*. These are very smooth and do not show any structure visible in the electron microscope. Only after extraction with such a strong leaching agent as 10 per cent KOH can the microfibrils be made visible. Therefore, the new wall formed at the tip seems to be cutinized instantly in order to prevent desiccation of the pollen tube which is growing in air. The microfibrils of pollen tubes have a smaller diameter than those of the microfibrils found hitherto in primary cell walls. The supposition that this might be due to the haploid state of the pollen tubes turned out to be wrong. There is no relationship between the size of microfibrils and the number of genoms in the nucleus.

This was shown by looking at the haploid alga *Spirogyra*, where very coarse microfibrils occur in the cell wall. Incidentally it was found that the rounded wall of the end cell in a *Spirogyra* thread showed an unexpected tip growth. At the very tip, longitudinal microfibrils spread out from the woven texture of the wall into the open, leaving a gap between them. Along the margin of this gap cross-fibrils are woven into the pattern of the cell wall. It seems that the cytoplasm of the cell tip is so to speak naked while forming the new wall.

CELLS OF MACERATED COLEOPTILES

The extension growth has been studied in the cells of macerated coleoptiles of oat and maize. This classical subject has yielded quite unexpected results. At a very young stage (oat coleoptiles of 5 mm. length) the *parenchyma cells* show an unexpectedly early formation of local secondary bands with parallel texture laid down on the woven primary wall. These reinforcements were found to be located along the edges of the polyhedral cells.

It is difficult to understand how such a parallel-textured strip can extend when the cell elongates. These strips are also visible in the polarizing microscope when stained with benzoazurin. Between the reinforced edges there are thin pitted areas. There again an extension seems quite impossible without distortion of the pits whose elliptic orifices have always a transverse orientation. The elongation of the cells is actually not due to a growth in area of the cell faces already formed, but to a polar or even bipolar tip growth.

Since cells are rare in the macerated tissue, this tip growth must go on very rapidly. A gap bordered by longitudinal microfibrils is clearly seen, as well as a margin where transverse microfibrils are woven into this warp. In the living state the tip seems to consist of cytoplasm only, so that the wallbuilding protoplasm protrudes beyond the formed solid wall. This observation excludes any active help of the turgor pressure in cell elongation, because this would force the cytoplasm out of the cell rather than extend the woven framework of the cell wall. Cells with tip growth burst at the very tip when their turgor pressure is increased.

This impressive phenomenon of plasmoptysis at the cell tip can easily be observed when growing pollen tubes are immersed in water. *Epidermal cells* prove still more conclusively that cell extension is really tip growth. In oat coleoptiles these cells elongate from about 13 to 2000 ¼ (2 mm.), *i.e.* 150 times within 4 days.

There is a very thick exterior wall which already exists before the cell extension starts and which holds the same thickness when the extension proceeds. The thick outer wall is visible as a broad parallel-textured band at the top of the figure, whilst the rest of the cell is tubular-textured as in the parenchyma cells. At the cell tip this tubular texture grows, and the same manner of growth is used for the elongation of the thick parallel-textured exterior wall.

There a much denser tuft of microfibrils is seen which is formed by the living cytoplasm from the cell. Thus the wall lengthens in its definite thickness by bipolar addition of parallelized microfibrils. *Sieve tubes* and *tracheids* also extend by tip growth.

For young fibres and tracheids this was already known. Whilst the cell has already differentiated to a ring tracheid at the bottom of the picture, the tip consists of a veil of interwoven cellulose microfibrils ending at the top in an open tuft of the longitudinal fibrils. It is clearly seen that the cytoplasm forms the warp first, and only afterwards is the weft plaited into the woven primary wall.

Finally, the transverse microfibrils exceed those of the longitudinal warp in number. By this type of growth, parenchyma cells, epidermal cells (apart from the outer wall), sieve tubes and the primary wall of tracheids obtain their *tubular texture* with transverse pits and the highest value of anisotropic properties, such as refraction, absorption, strength, etc., in the transverse direction as had been previously established by indirect methods such as double refraction and dichroism.

STRUCTURE OF PLANT CELLS

The typical cell of the higher plantsis a tiny compartment enclosed by a tough elastic wall.

The wall of cells consists of two major parts:

1. The middle lamella and
2. The primary wall.

In walls of many plant cells a third structural component, the secondary wall, is also present. Although some plant cells are known which do not have a well defined cell wall, this structure is so generally present in plant cells as to be considered one of their characteristic features. Lining the interior of the wall and occupying more or less of the cell cavity-nucleus is the protoplasm. The protoplasm of active cells is a transparent, slightly viscous, granular material that any conspicuous structural background. It is not homoge-neous, however, and vacucle contains a number of definite structures. One of these, the nucleus, is a denser body which is more or less pheroidal in shape and is separated from the remaining protoplasm by a definite membrane, the nuclear membrane.

Perspective view of a rounding the nucleus are:

- A clear palisade cell from the leaf mesophyll. liquid known as the nuclear sap,
- A delicate network of denser material, the reticulum, and
- One or more small spherical masses of material known as the nucleolus or nucleoli.

All of the protoplasm outside of the nucleus of the cell constitutes the cytoplasm. In a typical mature plant cell the cytoplasm is present as a thin layer lining the inner surface of the cell wall. The two boundary layers ofthe cytoplasm that in contact with the cell wall and that in contact with vacuole are called the cytoplasmic membranes. Imbedded in the cytoplasm are numerous well differentiated bodies known as plastids.

Plastids are specialized cytoplasmic structures which are usually centres of certain types of physiological activity. They are commonly classified on the basis of their colour into three groups: The leucoplasts which are colourless, the cizioroplasts which contain the green chlorophyll pigments (also yellow pigments), and the chromoplasts which contain red or yellow pigments. Chondriosomes, minute rod-like or granular bodies, are also found in the cytoplasm. The significance of these structures is not positively known, although a number of different roles have been ascribed to them.

Although the cell wall appears to imprison each protoplast, and to effectively isolate it from the protoplasm of adjoining cells, actually there is probably a continuation of protoplasm from cell to cell. By certain techniques it can be demonstrated that minute pores extend from cell to cell through the cell walls. These pores often contain cytoplasmic strands which connect the cytoplasms of adjacent cells. These strands are termed plasmodesms.

Livingston has demonstrated the occurrence of plasmodesms in the walls of cells from a number of different tissues of the tobacco plant, and it is generally supposed that they are of widespread if not universal occurrence inplant cell walls.

The bulk of the interior of mature plant cells is occupied by a single large cavity, the vacuole, which is filled with cell sap. The cell sap is composed of water in which a great variety of substances are dissolved or colloidally dispersed. There is no general agreement among cytologists regarding the exact classification of the parts of plant cells. The vacuole, for example, is frequently classified as a part of the protoplasm because it first appears as minute droplets in the protoplasm of very young cells. Physiologically, however, the vacuole of the mature cell is as distinct an entity as the protoplasm or cell wall and it is therefore considered as a separate part of the cell in this book.

The following classification includes the principal parts of a mature plant cell. A few of these parts, as for example the various kinds of plastids, do not occur in every cell. Middle lamella, Cell Wall, Primary Wall, Secondary Wall, The Mature Plant Cell, Protoplasm, Cytoplasm, Nucleus, Vacuole, Plasmodesms, Cytoplasmic membranes, Undifferentiated cytoplasm, leuco plasts, Plastids, chloroplasts, chromoplasts, Chondriosomes nuclear membrane, Nuclear sap Reticulum Nucleolus Water, Various compounds in solution or in a state of colloidal dispersion.

ORIGIN AND DEVELOPMENT OF PLANT

As soon as it became clear that all plant and animal tissues were composed of cells the question of the origin of cells naturally arose. This proved to be a difficult problem for the pioneer investigators and for many years there was much disagreement over the question. The now universally accepted principle that cells can arise only by the division of pre-existing cells was first demonstrated beyond any reasonable doubt by Niigeli just before the middle of the Nineteenth Century. A number of methods are now known by which this division is accomplished but a detailed discussion of them is beyond the scope of this book. In higher plants cell division occurs chiefly in certain restricted regions called meristems. The production of new cells involves not only the division of pre-existing cells, but the subsequent enlargement and maturation of their cell progeny.

FORMS AND SIZES

All newly formed cells do not differentiate morpholo-gically in the same way. Some elongate parallel to the axis of growth more than in other directions and thus produce the longer fibre cells that make up much of the xylem and phloem tissues. These cells commonly develop walls that are greatly thickened. Some cells enlarge about equally in all directions forming isodiarnetric cells, the walls of which are never greatly thickened. Cells of this type are present in the pith, leaf mesophyll and in other parenchymatous tissues.

According to Lewis both plant and animal cells are basically tetrakaidecahedrons; *i.e.*, 14 sided although many other geometrical shapes are found, especially in specialized types of cells. An almost endless variation in cell shapes and sizes may be seen in the tissues of any vascular plant, all of which may develop from similarly shaped meristematic cells. In size cells show an equally great range. Most plant cells have diameters that fall somewhere between 10 and 100. A single cubic centimeter oftissue may, therefore, contain millions of cells. Some cells, however, are much larger. Such cells have lengths that are many thousand times their diameters.

STRUCTURE OF CYTOPLASM

Any satisfactory explanation of the structure of cytoplasm must account not only for its physical properties and for its dynamic behaviour, but must also explain how the innumerable diverse physical and chemical reactions characteristic of living cells can occur side by side in the same general medium.

The highest magnifications reveal no evidence of any structural background in active cytoplasm, yet its complex activities suggest that it must possess an intricate structural organization. The marked imbibitional capacity of cytoplasm, its stability towards electrolytes, its electrical properties, its viscosity, its coagulability and its gel-forming capacity all suggest that it is to be classed with the hydrophilic colloids. All modern students of cytoplasm are agreed that it is a hydrophilic colloidal system of extreme complexity and variability.

As was pointed out in the preceding chapter elastic gels are generally considered to possess a submicroscopic structure of long interlacing fibrillar units which hold a liquid phase in their irregular interstices. The same structure has been suggested for some hydrophilic sols. There is some evidence that the structure of cytoplasm is similar. The cytoplasm in many living cells appears fibrous and long delicate strands may extend through the vacuole of the cell. Thread-like strands likewise appear in cytoplasm when it is being manipulated with very fine glass needles.

The sudden changes in viscosity that living cytoplasm may undergo when disturbed, strongly suggest a fundamental structure similar to that of elastic gels. Furthermore, living protoplasm can itself flow slowly through small openings such as those in a fine sieve without injury, but cannot be forced through somewhat larger holes without serious injury or death. This behaviour has been interpreted as evidence that a fibre-like structure exists in protoplasm; the fibres being able to slip through the holes when it is slowly streaming but being crushed and destroyed if a mechanical pressure is utilized to force the protoplasmic mass through the openings.

On the basis of this and other evidence living cytoplasm is presumed by some workers to have a skeletal structure of long submicroscopic fibrils. A fluid phase in which are suspended the visible droplets and granules is believed to be held in the interstices between the fibrils. Solutes may be presumed to be present in both phases. The whole system of fibrils and liquidis assumed to be dynamic and undergoing constant change, in which gel-like fibrils are rapidly transformed into sols and the sols form fibrils with equal rapidity. Such a structure will account satisfactorily for many of the properties of cytoplasm.For example, it offers an explanation for the rapid changes in viscosity that are known to occur in cytoplasm, it accounts satisfactorily for its hydrophilic properties, and it suggests a possible solution to the perplexing problem of how so many diverse processes can go on in the same cytoplasm at the same time. Many of these reactions may occur at the interfaces between the various fibrillar units and the liquid constituents of the system. Since different fibrillar units are undoubtedly of different chemical composition, different chemical and physical reactions could take place some at one interface and some at another. Furthermore, it is possible that some reactions occur within the liquid component and others within the more solid fibrillar units of the cytoplasm.

THE PLASTIDS

All living cells of the higher plants contain prominent cytoplasmic bodies known as plastids. These structures are usually ellipsoidal in shape and are frequently conspicuous because of the presence of pigments. The colour of the pigments in the plastids has been used as the basis for their classification but this system is highly artificial since a single plastid may be colourless, green, red or yellow at different periods of its existence.In meristematic and embryonic cells the plastids first appear as very tiny granules in the cytoplasm that range in size down to the limit of visibility. These granules gradually enlarge and differentiate untilmature plastids are produced. In growing and in mature cells plastids frequently multiply by simple fission.

In the algae the chioropiasts exhibit a wide range of size and shape but in the higher plants they show a remarkable similarity. The mature chioroplast of the higher plants is typically a slightly flattened ovate spheroid with the longer axis ranging between 4 and 6 in length. The number of plastids is not constant in different cells nor in the same cell at different stages of its development.

The structure of chloroplasts has been more thoroughly investigated than that of other kinds of plastids. Most investigations indicate that the chloroplast consists basically of a proteinaceous matrix or stroma which is probably surrounded by a membrane. In at least some chloroplasts the chlorophyll apparently occurs in small disk-shaped granules called grana which are distributed throughout the stroma. The chlorophyll is apparently associated with both proteins and lipids. The yellow pigments carotene and the xanthophylls also occur in the chloroplasts.

The chromoplasts contain red or yellow pigments and are frequently very different in size and shape from the chioroplasts. Usually they are very slender spindle-shaped or needle-shaped bodies. They occur both singly and grouped in bundles. The irregular angular outline of these plastids contrasts sharply with the regular curved surface of the chloroplasts. In some species however, chloroplasts may develop red or yellow pigments and comple-tely lose their green colour. The red and yellow colours of some fruits, notably members of the Solanaceae, and some flowers are caused by chromoplasts.

The colourless leucoplasts are usually present in the cells of meristematic tissues in which they often represent juvenile stages in the development of chloroplasts and chromoplasts. In tissues not exposed to light they remain as colourless plastids and are the structures in which starch grains are formed. Specialized plastids known as elaioplasts have been described as occurring in the cells of some species. These plastids appear to be centres of oil formation.

THE NUCLEUS

The nucleus is a conspicuous spheroidal body which is imbedded in the cytoplasm. In most plant cells the nucleus has a diameter which falls within a range of 5 to 25. In the vascular plants there is usually only one nucleus to a cell, although in certain types of cells several may be present. The sieve tube elements are the only well known example of living cells in the higher plants in which no organized nucleus is present.

The nucleus is surrounded by a definite membrane and possesses a complicated internal structural organization. The larger portion of the volume of the nucleus is composed of a transparent, optically homogeneous sol or gel the nuclear sap, which surrounds a system of delicate, anastomosing threads the reticulum. The reticulum is not structurally homogeneous but contains irregular granules of material known as chromatin. Most nuclei also contain one or more nucleoli. Typically a nucleolus is a small spherical droplet of deeply staining protein and lipoidal material that is attached to the reticulum.

The hereditary factors which influence the development of the organism are known to reside in the nucleus of the cells. It is clear therefore that the nucleus must exert a controlling influence over the physiological activities of the cell. There is some evidence that the nucleus is concerned with the production of the enzymes which catalyze many, if not most, physiological processes. Very little is known, however, regarding the exact role of the nucleus in protoplasmic activities.

THE VACUOLE

One of the most characteristic features of a mature plantcell is the presence of a large central vacuole filled with cell sap and entirely surrounded by the cytoplasm.

Merismatic cells in the tips of stems and roots usually contain numerous small vacuoles scattered throughout the cytoplasm. The shape of these vacuoles varies greatly and seems to be determined by the activity of the cytoplasm. In quiescent cytoplasm the small vacuoles are usually spherical but when the cytoplasm is actively streaming they may assume many different forms.

Cambium cells that are actively dividing may contain very large vacuoles. In cambium initials the vacuoles, like those in the meristems of stem and root tips, are not uniform in size or shape, but may berod-shaped, thread-like or globular. They may coalesce into a large single vacuole or they may divide into numerous smaller vacuoles. Mature cells however, whether they arise from primary meristems or from the cambium cells typically contain one large central vacuole which arises by the increase in size and coalescence of the numerous smaller vacuoles usually present in the meristematic cells. There is no general agreement as to the method by which vacuoles originate.

Three possibilities are recognized:

1. They may arise by the division of pre-existing vacuoles,
2. They may originate in the cytoplasm, and
3. They may develop from organized units of the cytoplasm. There is no convincing evidence, however, that the vacuoles of the cells of the vascular plants arise in any way except by the division of pre-existing vacuoles. Among the various substances present as solutes in the vacuole are sugars mineral salts, organic acids (oxalic acid, especially, seems to be of frequent occurrence), amino acids, amides, alkaloids, glycosides, flavones, and anthocyanin. Fats and related compounds often occur in finely emulsified form. Proteins, tannins, mucilages, lipids and other substances are commonly present in the colloidal state. Aleurone grains develop from specialized vacuoles of cells in storage tissues. Crystals of calcium oxalate are also of frequent occurrence in the vacuoles of mature cells.

CONCENTRATION OF PLANT CELLS

The hydration and viscosity of the protoplasm, the permeability of the cytoplasmic membranes, the activity of enzymes, the chemical activity of various ions in the cell, and various other physiological processes and conditions are all influenced more or less by the hydrogen ion concentration of the protoplasm and the cell sap. As soon as its significance was appreciated attempts were made to determine the hydrogen ion concentration of plant cells. Most of the earlier determinations were made on the juice pressed from plant tissues. Such a crude method provides only the roughest sort of an indication of the hydrogen ion concentrations in individual plant cells. The death of the cells and the mixing of the cell contents during the extraction process undoubtedly result in marked changes in hydrogen ion concentration. Such determinations are probably more nearly a measure of the pH of the cell sap than of the protoplasm. The values for expressed plant saps mostly fall within a range of pH 3.Q to pH 7.0 Direct measurements of the hydrogen ion concentration of the protoplasm and cell sap have been made by introducing indicator dyes directly into cells.By careful manipulation of a micropipette the dyes can be injected into the cytoplasm without penetrating into the vacuole or injected into the vacuole without penetrating into the cytoplasm. Only non-toxic dyes should be injected into living cytoplasm for pH determinations, else the results may be invalidated by injury or death of the cytoplasm. Dyes are considered to be nontoxic or essentially so if protoplasmic streaming continues in the same way as before the injection. In this manner it is possible to determine the reaction of the cell sap and that of the cytoplasm independently. Results of the micro injection method when applied to the root hairs of the water plant Limnobiumspongia indicated a pH value for the cytoplasm of 6.9 1 0.2. The cell sap of the same cells was found to be more acid having a pH of 5.2 1 0.2.

As indicated by the results cited above the pH of the cytoplasm and the cell sap of a plant cell may be very different. The pH of the cytoplasm of plant cells appears to be fairly constant, usually falling between 6.8 and 7.0. The pH of the cell sap is usually lower than that of the cytoplasm, values between pH 5.2 and 6.2 seeming typical for most plant cells.The cell sap of some cells, however, shows a considerably more acid reaction than this, values as low as p11 0.9 being reported for species of Begonia. On the other hand alkaline values have also been found in the vacuoles of some species. The cells appears to vary more widely in hydrogen ion concentration than the cytoplasm.

CELL-WALL GROWTH IN DIVIDING CELLS

How does the cell wall behave when the cell divides? In very young coleoptiles there is still some mitotic activity. We could therefore photograph a certain number of dividing cells. The structure of the expanding cell plate is difficult to observe, because it is inside the enveloping membrane of the mother cell. But the constriction of this wall in the middle of the cell is clearly visible. Its microfibrils are partly bent inwards, where they meet the cell plate. Thereupon they are interwoven with the microfibrils of the corresponding lamella of the double-laminated cell plate. A certain number of the microfibrils of the original longitudinal wall are left over and decorate the margin of the two daughter cells with their submicroscopic fringes. These microfibrils are gradually dissolved and disappear. This proves that existing cellulose fibrils can be hydrolysed near by the place where other microfibrils are synthesized. Whether these two antagonistic processes are catalysed by the same enzymatic system is open to discussion.

The effort of our laboratory to elucidate the mechanism of extension growth has shown that, in the object investigated, there is no difference between tip growth and 'cell extension'. This had already been postulated previously. But it was thought then that the primary wall could expand by loosening its existing framework and subsequently inserting new microfibrils. In order to obtain a local loosening, certain microfibrils had to be dissolved by hydrolysis.

That a local 'melting down' of microfibrils is possible has been proved in this study, when we observed how the ring of microfibrils left over as fringes along the border of the new wall in divided cells gradually disappears. Thus, plasticizing of an existing cellulose frame by dissolution of microfibrils is possible. But in the observed cases of cell extension it does obviously not occur. Growth in area does not proceed by intussusception but by a terminal increment, just as the area of a cloth is enlarged by weaving its margin. In so doing the cytoplasm must somehow protrude beyond its solid case.

As the coleoptiles investigated represent the classical subject for the study of extension growth, there is some danger of generalizing these findings to all cases of growth in area of cell walls and of banishing the concept of intussusception. Although most elongating cells probably grow in the way here described, there are still some cases which must be thoroughly investigated before a general theory of cell-wall growth can be formulated. It will be very interesting to find out how tip growth proceeds in the very fast extension growth of grass filaments.

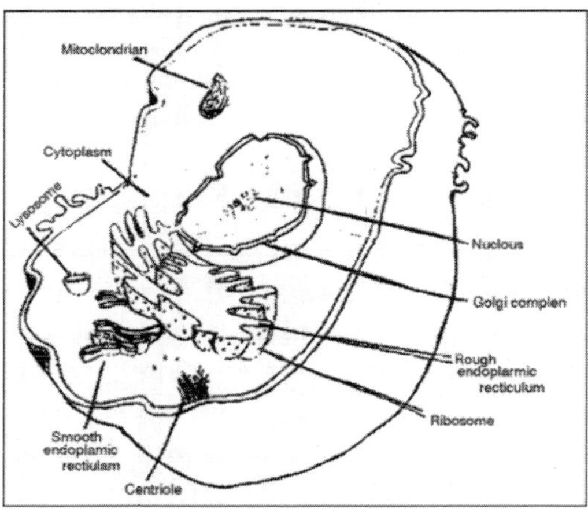

Fig. Cytoplasm

Further, there is an enormous growth in area of the wall when the descendant of a narrow cambial cell differentiates to form a large-bored vessel member. Here no tip growth is possible, but the cell surface must increase along some longitudinal line or lines, where plasticizing of the existing wall is necessary. The same is true when a cell produces some local outgrowth (*e.g.* a root hair). At the place where this reactivated tip growth starts, the existing framework must have been loosened somehow beforehand.

Since the texture of primary walls is so spacious, there are many possibilities of rendering the wall softer by increasing the interfibrillar ('incrusting') substances capable of swelling, or by intercalating new microfibrils. This would be possible where living cytoplasm is located within the growing cell wall. Our study seems to indicate, however, that the cell wall is only soaked with living matter locally. Thus intussusception is restricted to small areas, and the cytoplasm retires very soon from the woven framework. All these considerations show that the growth of the plant cell wall is an autonomic process. Exterior forces such as turgor pressure cannot play an essential role;

they are only involved in so far as they influence the activity of the cytoplasm. The extension growth of the cells in coleoptiles is a *bipolar tip growth*. The growing cell wall is living: it must be soaked with cytoplasm which synthesizes microfibrils *in situ*. In dividing cells it bends them to meet the cell plate on which they are fastened. Therefore the growth of the cell wall is an autonomic process which is governed by the morphogenetic faculties of the living protoplast.

3

Plant Tissue Culture and Biotechnology

PLANT TISSUE CULTURE

Plant tissue culture is a method or technique to isolate parts of plants (protoplasm, cells, tissues, and organs) and grow them on artificial media in aseptic conditions in a controlled space so that parts of these plants can grow and develop into complete plants. Plant tissue culture techniques are essential to many types of academic inquiry, as well as to many applied aspects of plant science. In the past, plant tissue culture techniques have been used in academic investigations of totipotency and the roles of hormones in cytodifferentiation and organogenesis.

Currently, tissue-cultured plants that have been genetically engineered provide insight into plant molecular biology and gene regulation. Plant tissue culture techniques are also central to innovative areas of applied plant science, including plant biotechnology and agriculture. For example, select plants can be cloned and cultured as suspended cells from which plant products can be harvested. In addition, the management of genetically engineered cells to form transgenic whole plants requires tissue culture procedures; tissue culture methods are also required in the formation of somatic haploid embryos from which homozygous plants can be generated. Thus, tissue culture techniques have been, and still are, prominent in academic and applied plant science.

The techniques demonstrated in these exercises range from simple ones that can easily be performed by beginning students to those done by botany or physiology students. The use of tissue culture techniques in the beginning just to prove the theory of "totipotensi" ("total genetic potential") is expressed by Schleiden and Schwann which states that the plant cell as the smallest unit can grow and thrive if kept in appropriate conditions. We have used tissue culture techniques not only as a means for studying aspects of plant physiology and biochemistry, but has developed into methods for various purposes.

TISSUE CULTURE TECHNIQUES

An important aspect of all biotechnology processes is the culture of either the plant cells or animal cells or microorganisms. The cells in culture can be used for recombinant DNA technology, genetic manipulations etc.

Plant cell culture is based on the unique property of the cell-totipotency. CELL-TOTIPOTENCY is the ability of the plant cell to regenerate into whole plant. This property of the plant cells has been exploited to regenerate plant cells under the laboratory conditions using artificial nutrient mediums. With the advances made in genetic engineering, it became possible to introduce foreign genes into cell and tissue culture systems. This led to the development of Genetically Modified (Gm) or Transgenic Crops which had improved traits and characteristics.

HISTORY OF CELL CULTURE

In the early 19th century, Schleiden and Schwann proposed the concept of the 'cell theory'. In 1902, Gottlieb Haberlandt, the german botanist and regarded as the father of plant tissue culture, first attempted to cultivate the mechanically isolated plant leaf cells on a simple nutrient medium. He did not succeed in achieving the growth and differentiation of the cultured cells, however, he predicted the concept of growth hormones, the use of embryo sac fluids, the cultivation of artificial embryos from somatic cells, etc.

During the period 1902 - 1930, attempts were made to culture the isolated plant organs such as roots and shoot apices (organ culture). Hanning isolated embryos of some crucifers and successfully grew on mineral salts and sugar solutions. Simon successfully regenerated a bulky callus, buds, roots from a poplar tree on the surface of medium containing IAA which proliferated cell division. Gautheret, White and Nobecourt largely contributed to the developments made in plant tissue culture. White cultured tobacco tumour tissue from the hybrid Nicotiana glauca, and N. Langsdorffii.

The period saw the development of suitable nutrient media to culture plant tissues, embryos, anthers, pollen, cells and protoplasts, and the regeneration of complete plants (in vitro morphogenesis) from cultured tissues and cells. In 1941, van Overbek and co-workers used coconut milk (embryo sac fluid) for embryo development and callus formation in Datura. Steward and Reinert first discovered somatic embryo production in vitro. Maheswari and Guha developed the anther culture for the production of haplid plants. Skoog and Miller advanced the hypothesis of organogenesis in cultured callus by varying the ratio of auxin and cytokinin in the growth medium. Muir developed a successful technique for the culture of single isolated cells wich is commonly known as paper-raft nurse technique (placing a single cell on filter paper kept on an actively growing nurse tissue). In 1952, the Pfizer Inc., New York got the US patent and started producing industrially the secondary metabolites of plants. The first commercial production of a natural product shikonin by cell suspension culture was obtained.In 1980s using Genetic engineering, for the first time, it was possible to introduce foreign genes into cell and tissue culture systems to develop plants with improved characteristics (transgenic crops) which may contribute to the path towards the second green revolution.A transgenic animal is an individual in which a gene (or genes) from another individual has been artificially inserted with the new genetic information forming a permanent part of the genome and being passed on to their offspring according to the Mendelian laws of inheritance.

So far, transgenic cattle, sheep, pigs, rabbits, and chicken have been produced.After fertilization a mammalian embryo passes through many stages of development such as formation of pronuclei, morula formation, blastocyst formation, implantation etc. Using these early development stages of mammalian embryos, genetic engineers have used different methods to introduce foreign genes into the genome.

Table. List of Transgenic Animals Produced With Their Promoter, Enhancer and Structural Transgenes

Animal	Genes Tansferred (Promoter or Enhancer/Structural Transgenes)
Goat	A variant of tPA gene (LAtPA)
Sheep	mMT/hGH, mMT/TK, mMT/bGH, mMT/hGRF, oBLG/hFIX, oBLG/alpha1AT, oMT/oGH
Rabbit	mMT/hGH, hMT/hGH, rbEu/rb,c-myc
Pig	mMT/hGH, mMT/bGH, hMT/pGH, MLV/rGH, MLV/rGH, bPRL/bGH
Fish	hGH, mMT/hGh, mMT/bGal, cd-crystallin SV/hygro, AFP
Cow	BPV, lactoferrin
Chicken	ALV, REV
Mouse	mMT/rGH, mMT/bGH, mMT/oGH, mMT/hGh, mMT/hGRF, mMT/hFIX

METHODS TO INDUCE FOREIGN GENES INTO THE GENOME

Micro-injection in to a Pronucleus

A fertilized egg is held in a pipette by suction and several copies of the foreign genes are injected into one of the pronuclei via a micropipette. Although this has been the most widely used and most successful method, 30% of the treated embryos degenerate and die with in a few hours.

Using Retroviruses as Vectors to CarryForeign DNA into Morulas

A morula is placed on a culture of fibroblasts infected with retrovirus after removing the zona pellucida layer. The virus cannot penetrate this layer therefore it is essential to get rid of it. The retrovirus, genetically engineered to carry foreign DNA, infect the cells of the morula as they are shed from the fibroblast. This technique often creates 'mosaic offspring', where some cells contain the foreign gene while others do not. One of the limitations of this technique is that the modified virus cannot leave the transgenic species and under certain conditions it could mutate regaining its ability to cause disease in the tissues where it occurs.

Retroviral Infection of the Stem cells, which are then Injected into the Cavity of Another Blastocyst

In this method, stem cells, from the inner cell mass of an embryo, are infected with genetically-engineered retroviruses, then injected into the central cavity of a different blastocyst. The injected cells colonise the new embryo and participate in the formation of all the tissues, including the ovaries and testes from which cells are formed. This technique has been used to produce transgenic mice and hamsters. The first transgenic animals produced were mice, into which a gene cloning for rat growth hormone was inserted by microinjection.

Two techniques were used to introduce the new gene into the mouse genome:

- As many as 10000 copies of the rat gene were injected into a pronucleus of a fertilized mouse egg, using a microneedle.
- After treatment with calcium chloride, to make them more permeable, the rat DNA was applied to the outside of a 2-8- celled mouse embryo in culture.

The treated eggs, or embryos were then transplanted into foster mothers. Embryos with the inserted genes were identified by DNA hybridization with small samples of DNA from white blood cells or skin cells. Transformed adult mice produced the growth hormone mainly in their livers rather than pituitary gland where it is normally synthesized.This technique can be used successfully to transfer other genes such as those affecting the properties of hair, hides, cold tolerance, disease-resistance and milk production which might help to produce strains of farm animals with economic advantage over existing breeds and stocks. In this regard, research is going on to use techniques which can improve the quality of the farm animals. In this regard, following two techniques are of great importance.

IMPROVING THE QUALITY

EMBRYO MANIPULATION

In this method, the fertilized egg is bisected at the two celled stage and each half is transplanted into different regions of the uterus. Using this method the reproductive rate is doubled as we know that female sheep and cattle produce, on an average, one offspring per pregnancy. Hence the farmers can easily increase their farmstock.

This technique can also be used to conserve rare breeds. After fertilizing their eggs in the laboratory, the young embryos or rare animals are dissected into anything from 2-8 cells. Each cell, is then transplanted into a surrogate mother of a common breed where it grows to produce a new individual of the rare type.

Another variation of this technique allows surrogate mothers to carry embryos of a different species *e.g.* Horses giving birth to zebras from zebra embryos implanted at the blastocyst stage. The outer layer of zebra embryo is exchanged with the trophectoderm of the horse embryo in order to make it acceptable to the surrogate mother. This exchange of the layer does not affect the development of the inner cell mass which is the true embryo-forming region.

Embryo Cloning

Using this technique it is possible to produce many genetically identical copies of an animal. This method has been used to clone mouse however, research is going on to standardize this method to obtain clones of cattle with desirable traits *e.g.* cows with high milk production etc.

Following steps are used in this method:

- An egg from the donor is grown to blastocyst stage under laboratory conditions.
- The egg is dissected to remove the inner cell mass.
- The mass cells are separated into individual cells.
- Injection of a nucleus from each of these cells into a one-celled embryo containing two pronuclei.
- Removal of pronuclei followed by culturing of embryos in the laboratory until the blastocyst stage.
- Transplantation of these embryos into surrogate mothers.

CHIMERAS

Early research into the production of transgenic animals revealed a simple method for producing hybrids, called Chimeras, between closely related species. Sheep-goat chimeras were produced by mixing four celled sheep embryos with eight-celled goat embryos. After removing the zona pellucida from each egg, the eggs were pressed together and incubated at 370C. The cells reorganized and formed hybrid blastocysts. These hybrid blastocysts were then transferred to sheep foster mothers, where they continued their growth and development until the sheep gave birth. Each hybrid offspring contained both sheep and goat cells in all it's tissues which resulted in the coats of these animal having a patchwork kind of pattern with irregular patches of sheep and goat fur.

ENVIRONMENTAL TECHNIQUES

The term "Environment" is defined as our surroundings which includes the abiotic component (the non living) and biotic component (the living) around us. The abiotic environment includes water, air and soil while the biotic environment consists of all living organisms – plants, animals and microorganisms. Environmental pollution broadly refers to the presence of undesirable substances in the environment which are harmful to man and other organisms. In the past decade or two, there has been a significant increase in the levels of environmental pollution mostly due to direct or indirect human activities. The major sources of environmental pollution are –Industries, Agricultural sources (mainly rural area), anthropogenic sources (man related activities mainly in urban areas), biogenic sources etc. The pollutants are chemical, biological and physical in nature. The Chemical pollutants include- gaseous pollutants (hazardous gases like sulfur dioxide, nitrogen oxide), toxic metals, pesticides, herbicides toxins and carcinogens

Etc. The physical pollutants are- heat, sound, radiation, and radioactive substances. The pathogenic organisms and some poisonous and dangerous biological products are the biological pollutants.

Controlling the environmental pollution and the conservation of environment and biodiversity and controlling environmental pollution are the major focus areas of all the countries around the world. In this context the importance and impact of biotechnological approaches and the implications of biotechnology has to be thoroughly evaluated. There have been serious concerns regarding the use of biotechnology products and the impact assessment of these products due to their interaction with the environmental factors.

A lobby of the environmentalists have expressed alarm on the release of genetically engineered organisms in the atmosphere and have stressed on thorough investigation and proper risk assessment of theses organisms before releasing them in to the environment. The effect of the effluents from biotechnological companies is also a cause of concern for everyone. The need of the hour is to have a proper debate on the safety of the use of the biotechnological products.The efforts are not only on to use biotechnology to protect the environment from pollutionbut also to use it to conserve the natural resources. As we all know that microorganisms are known natural scavengers so the microbial preparations (both natural as well as genetically engineered) can be used to clean up the environmental hazards.

DEVELOPMENT OF ALTERNATE CLEANER TECHNOLOGIESUSING BIOTECHNOLOGY

Biotechnology is being used to provide alternative cleaner technologies which will help to further reduce the hazardous environmental implications of the traditional technologies. *E.g.* some Fermentation technologies have some serious environmental implications. Various biotechnological processes have been devised in which all nutrients introduced for fermentation are retained in the final product, which ensures high conversion efficiency and low environmental impact.In paper industry, the pulp bleaching technologies are being replaced by more environmentally friendly technologies involving biotechnology.

The pulp processing helps to remove the lignin without damaging valuable cellulosic fibres but the available techniques suffer from the disadvantages of high costs, high energy use and corrosion. A lignin degrading and modifying enzyme (LDM) was isolated from Phanerochaete chrysosporum and was used, which on one hand, helped to reduce the energy costs and corrosion and on the other hand increased the life of the system. This approach helped in reducing the environmental hazards associated with bleach plant effluents.

In Plastic industry, the conventional technologies use oil based raw materials to extract ethylene and propylene which are converted to alkene oxides and then polymerized to form plastics such as polypropylene and polyethylene. There is always the risk of these raw materials escaping into the atmosphere thereby causing pollution. Using biotechnology, more safer raw materials like sugars (glucose) are being used which are enzymatically or through the direct use of microbes converted into alkene oxides.*e.g.* Methylococcus capsulatus has been used for converting alkene into alkene oxides.

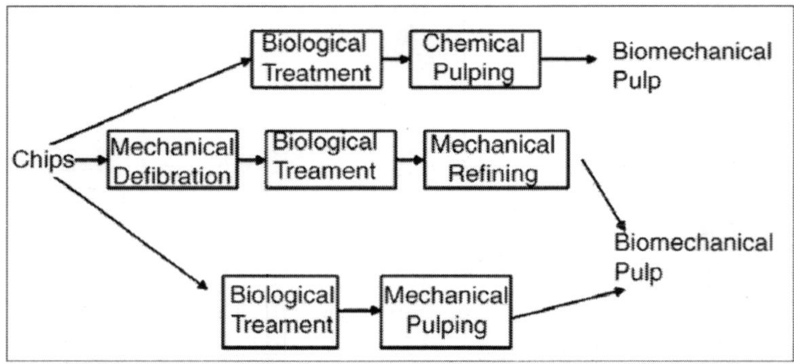

Fig. Integration of Biological Steps in Pulping ProcessLeading to Lignin Degradation

BIOREMEDIATION

Bioremediation is defined as 'the process of using microorganisms to remove the environmental pollutants where microbes serve as scavengers. The removal of organic wastes by microbes leads to environmental cleanup. The other names/terms used for bioremediation are biotreatment, bioreclamation, and biorestoration. The term "Xenobiotics" (xenos means foreign) refers to the unnatural, foreign and synthetic chemicals such as pesticides, herbicides, refrigerants, solvents and other organic compounds. The microbial degradation of xenobiotics also helps in reducing the environmental pollution. Pseudomonas which is a soil microorganism effectively degrades xenobiotics. Different strains of Pseudomonas that are capable of detoxifying more than 100 organic compounds (*e.g.* phenols, biphenyls, organophosphates, naphthalene etc.) have been identified. Some other microbial strains are also known to have the capacity to degrade xenobiotics such as Mycobacterium, Alcaligenes, Norcardia etc.

FACTORS AFFECTING BIODEGRADATION

The factors that affect the biodegradation are: the chemical nature of xenobiotics, the concentration and supply of nutrients and O_2, temperature, pH, redox potential and the capability of the individual microorganism. The chemical nature of xenobiotics is very important because it was found out that the presence of halogens *e.g.* in aromatic compounds inhibits biodegradation. The water soluble compounds are more easily degradable whereas the presence of cyclic ring structure and the length chains or branches decrease the efficiency of biodegradation. The aliphatic compounds are more easily degraded than the aromatic ones.

BIOSTIMULATION

It is a process by which the microbial activity can be enhanced by increased supply of nutrients or by addition of certain stimulating agents like electron acceptors, surfactants etc.

BIOAUGMENTATION

It is possible to increase biodegradation through manipulation of genes *i.e.* using genetically engineered microorganisms and by using a range of microorganisms in biodegradation reaction.

Depending on the method followed to clean up the environment, the bioremediation is carried out in two ways:

In Situ Bioremediation

In situ bioremediation involves a direct approach for the microbial degradation of xenobiotics at the site of pollution which could be soil, water etc. The adequate amount of essential nutrients is supplied at the site which promotes the microbial growth at the site itself. The in situ bioremediation is generally used for clean up of oil spillages, beaches etc.

There are two types of in situ bioremediation:

1. *Intrinsic bioremediation*: The microorganisms which are used for biodegradation are tested for the natural capability to bring about biodegradation. So the inherent metabolic ability of the microorganisms to degrade certain pollutants is the intrinsic bioremediation. The ability of surface bacteria to degrade a given mixture of pollutants in ground water is dependent on the type and concentration of compounds, electron acceptor and the duration of bacteria exposed to contamination. Therefore, the ability of indigenous bacteria degrading contaminants can be

determined I laboratory by using the techniques of plate count and microcosm studies. The conditions of site that favour intrinsic bioremediation are ground water flow throughout the year carbonate minerals to buffer acidity produced during biodegradation, supply of electron acceptors and nutrients for microbial growth and absence of toxic compounds.

2. *Engineered in situ bioremediation*: When the bioremediation process is engineered to increase the metabolic degradation efficiency (of pollutants) it is called engineered in situ bioremediation. This is done by supplying sufficient amount of nutrients and oxygen supply, adding electron acceptors and maintaining optimal temperature and pH. This is done to overcome the slow and limited bioremediation capability of microorganisms.

Advantages of in Situ Bioremediation

- The method ensures minimal exposure to public or site personnels.
- There is limited or minimal disruption to the site of bioremediation.
- Due to these factors it is cost effective.
- The simultaneous treatment of contaminated soil and water is possible.

Disadvantages of in Situ Bioremediation

- The sites are directly exposed to environmental factors like temperature, oxygen supply etc.
- The seasonal variation of microbial activity exists.
- Problematic application of treatment additives like nutrients, surfactants, oxygen etc.
- It is a very tedious and time consuming process.

Ex-situ Bioremediation

In this the waste and the toxic material is collected from the polluted sites and the selected range of microorganisms carry out the bioremediation at designed place. This process is an improved method over the in situ bioremediation method.

On the basis of phases of contaminated materials under treatment ex-situ bioremediation is classified into two:

1. *Solid phase treatment*: This system includes land treatment and soil piles comprising of organic wastes like leaves, animal manures, agricultural wastes, domestic and industrial wastes, sewage sludge, and municipal solid wastes. The traditional clean-up practice involves the informal processing of the organic materials and production of composts which may be used as soil amendment. Composting is a self heating, substrate-dense, managed microbial system which is used

to treat large amount of contaminated solid material. Composting can be done in open system *i.e.* land treatment and/or in closed treatment system. The hazardous compounds reported to disappear through composting includes aliphatic and aromatic hydrocarbons and certain halogenated compounds. The possible routes leading to the disappearance of hazardous compounds include volatilization, assimilation, adsorption, polymerization and leaching.

2. *Slurry phase treatment*: This is a triphasic treatment system involving three major components- water, suspended particulate matter and air. Here water serves as suspending medium where nutrients, trace elements, pH adjustment chemicals and desorbed contaminants are dissolved. Suspended particulate matter includes a biologically inert substratum consisting of contaminants and biomass attached to soil matrix or free in suspending medium. The contaminated solid materials, microorganisms and water formulated into slurry are brought within a bioreactor *i.e.* fermenter. Biologically there are three types of slurry-phase bioreactors: aerated lagoons, low shear airlift reactor, and fluidized-bed soil reactor. The first two types are in use of full scale bioremediation, while the third one is in developmental stage.

Advantages of Ex-situ Bioremediation

- As the time required is short, it is a more efficient process.
- It can be controlled in a much better way.
- The process can be improved by enrichment with desired and more efficient microorganisms.

Disadvantages of Ex-situ Bioremediation

- The sites of pollution remain highly disturbed.
- Once the process is complete, the degraded waste disposal becomes a major problem.
- It is a costly process.

Several types of reactions occur during the bioremediation/microbial degradation:

- *Aerobic bioremediation*: When the biodegradation requires oxygen O_2 for the oxidation of organic compounds, it is called aerobic bioremediation. Enzymes like monooxygenases and dioxygenases are involved and act on aliphatic and aromatic compounds.
- *Anaerobic bioremediation*:This does not require oxygen O_2. the degradation process is slow but more cost effective since continuous supply of oxygen is not required.
- *Sequential bioremediation*: Some of the xenobiotic degradation requires

both aerobic as well as anaerobic processes which very effectively reduces the toxicity *e.g.* tetrachloromethane and tetrachloroethane undergo sequential degradation.

Use of genetic engineering and genetic manipulations for more efficient bioremediation. In recent years, efforts have been made to create genetically engineered microorganisms (GEMs) to enhance bioremediation. This is done to overcome some of the limitations and problems in bioremediation.

These problems are:

- Sometimes the growth of microorganisms gets inhibited or reduced by the xenobiotics.
- No single naturally occurring microorganisms has the capability of degrading all the xenobiotics present in the environmental pollution.
- The microbial degradation is a very slow process.
- Sometimes certain xenobiotics get adsorbed on to the particulate matter of soil and thus become unavailable for microbial degradation.

As the majority of genes responsible for the synthesis of enzymes with biodegradation capability are located on the plasmids, the genetic manipulations of plasmids can lead to the creation of new strains of bacteria with different degradative pathways. In 1970s, Chakrabarty and his team of co-workers reported the development of a new strain of bacterium Pseudomonas by manipulations of plasmid transfer which they named as "superbug". This superbug had the capability of degrading a number of hydrocarbons of petroleum simultaneouslysuch as camphor, octane, xylene, naphthalene etc. In 1980, United States granted the patent to this superbug making it the first genetically engineered microorganism to be patented.

In certain cases, the process of plasmid transfer was used. *E.g.* The bacterium containing CAM (camphor degrading) plasmid was conjugated with another bacterium with OCT (octane degrading) plasmid. Due to non-compatibility, these plasmids cannot coexist in the same bacterium. However, due to the presence of homologous regions of DNA, recombination occurs between these two plasmids which results in a single CAM-OCT plasmid giving the bacterium the capacity to degrade both camphor as well as octane.A new strain of Pseudomonas sp. (strain ATCC 1915) has been developed for the degradation of vanillate (which is a waste product from paper industry) and sodium dodecyl sulfate (SDS, a compound used in detergents).

ASEPTIC TECHNIQUE

The essence of aseptic technique is the exclusion of invading microorganisms during experimental procedures. If sterile tissues are available, then the exclusion of microorganisms is accomplished by using sterile instruments and culture media concurrently with standard bacteriological transfer procedures to avoid extraneous contamination.

Media and apparatus are rendered sterile by autoclaving at 15 lbs/inch 2 (121°C) for 15 minutes. The use of disposable sterile plasticware reduces the need for some autoclaving. Alternative sterilization techniques such as filter sterilization must be employed for heat-labile substances like cytokinins. Aseptic transfers can be made on the laboratory bench top by using standard bacteriological techniques (*i.e.*, flaming instruments prior to use and flaming the opening of receiving vessels prior to transfer). Aseptic transfers are more easily performed in a transfer chamber such as a laminar flow hood, which is also preferably equipped with a bunsen burner.

If experimental tissues are not aseptic, then surface sterilization procedures specific to the tissues are employed. Common sterilants are ethyl alcohol and/or chlorox with an added surfactant. Concentration of sterilants and exposure time are determined empirically. When cultured *in vitro*, all the needs of the plant cells, both chemical and physical, have to met by the culture vessel, the growth medium, and the external environment (light, temperature, etc.). The growth medium has to supply all the essential mineral ions required for growth and development. In many cases (as the biosynthetic capability of cells cultured *in vitro* may not replicate that of the parent plant), it must also supply additional organic supplements such as amino acids and vitamins. Many plant cell cultures, as they are not photosynthetic, also require the addition of a fixed carbon source in the form of a sugar (most often sucrose). One other vital component that must also be supplied is water, the principal biological solvent. Physical factors, such as temperature, pH, the gaseous environment, light (quality and duration), and osmotic pressure, also have to be maintained within acceptable limits.

MICROBIAL

Plant cell culture is viewed as a potential means of producing useful plant products such that conventional agriculture, with all its attendant problems and variables, can be circumvented. These problems include: environmental factors (drought, floods, etc.), disease, political and labour instabilities in the producing countries (often Third World countries), uncontrollable variations in the crop quality, inability of authorities to prevent crop adulteration, losses in storage and handling. Thus, the production of useful and valuable secondary metabolites in large bioreactors located in the consuming country is an attractive proposal. Additional advantages of such processes include: controlled production according to demand and a reduced and requirement.

However, this technology is still being developed and despite the advantages outlined above, there are a variety of problems to be overcome before it can be adopted on a wide scale for the production of useful plant secondary metabolites. In theory, it is anticipated that such large scale suspension cultures will be suitable for industrial production of useful plant chemicals such as pharmaceuticals and food additives, in a manner similar to that of microbial fermentation.

Nevertheless, there are some significant differences between microbial and plant cell cultures that must be considered when attempting to apply plant cell cultures to the available technology. A comparison of some of the characteristics of plant and microbial cultures of relevance to fermentation.

To demonstrate some of the problems that can be encountered with plant cell cultures. The sensitivity to shear is due both to the large size of the cells and to the relatively inflexible cellulose cell wall. Thus, with normal blade impellers the cells may twist which will inhibit mitoses and, for this reason, air-lift fermentors are recommended by some researchers. The large size of the plant cell contributes to its comparatively high doubling time(12 h - several days), which thus prolongs the time required for a successful fermentation run.

The vacuole is the major site of product accumulation, and since product secretion is uncommon, the high metabolite yields seen in microorganisms that secrete product (thereby removing product inhibition of biosynthesis) cannot be expected. There is some ongoing research on membrane permeabilization of plant cells which may serve to relieve the constraints of product inhibition by facilitation of leakage into the extracellular medium. If this would also permit recycling of the biomass (*e.g.* via immobilization) it would help reduce production costs.

PLANT TISSUE CULTURE MEDIA

Plant Tissue Culture refers to the technique of growing plant cells, tissues, organs, seeds or other plant parts in a sterile environment on a nutrient medium. In the plant tissue culture process, plant parts are cut into small pieces that will give rise to many more individual plants. Soil or hydroponically grown plants only need fertilizer (a source of K, N and P plus trace minerals), water, air and light to grow, because they can make their own sugar, amino acids, etc.

Plant tissue cultures need an outside source of sucrose because the tissue culture cells go into shock when the explants are made. It helps to add vitamins in addition to the fertilizer and sucrose. Pre-mixed media can be purchased in packs for preparing 1 liter. One liter usually provides enough medium to fill 60 50-ml tubes with 15 ml. Some folks like to use deep Petri plates (20 × 100mm). These need about 25 ml of medium per plate.

Is sugar in the pre-mix? Generally the pre-mixed formulations DO NOT contain sugar, nor do they contain agar. A simple way to figure out if the formulation packet contains sugar is that the total weight on a 1 liter packet would be around 30 grams. If sugar is NOT included, the total weight will be less than 5 grams. Generally, folks add 20-30 grams of sucrose to plant tissue culture medium.

What agar to use? Agar is left out because these pre-mixes are often used for suspension cultures. Note that Nutrient Agar contains lots of food for bacteria in addition to the inert agar. Be sure to use AGAR (not nutrient agar). Gel-Rite and Phytagar are more pure preparations used by professionals. Gel-Rite is used at lower concentrations (grams per liter or%) than agar. Bacteriological agar works fine for schools. Typically, agar is added at 0.7%-0.8% or 7-8 grams/ liter.

PREPARATION OF PLANT TISSUE CULTURE MEDIUM

- Measure approximately 90% of the required volume of the deionized-distilled water in a flask/container of double the size of the required volume.
- Add the dehydrated medium into the water and stir to dissolve the medium completely. Gentle heating of the solution may be required to bring powder into solution.
- Add desired heat stable supplements to the medium solution.
- Add additional deionized-distilled water to the medium solution to obtain the final required volume.
- Set the desired pH with NaOH or HCL
- Dispense the medium into culture vessels.
- Sterilize the medium by autoclaving at 15 psi (121°C) for appropriate time period. Higher temperature may result in poor cell growth.
- Add heat labile supplements after autoclaving.

The 30 major soil groups of the WRB are *Acrisols, Albeluvisols, Alisols, Andosols, Anthrosols, Arenosols, Calcisols, Cambisols, Chernozems, Cryosols, Durisols, Ferralsols, Fluvisols, Gleysols, Gypsisols, Histosols, Kastanozems, Leptosols, Lixisols, Luvisols, Nitisols, Phaeozems, Planosols, Plinthosols, Podzols, Regosols, Solonchaks, Solonetz, Umbrisols*, and *Vertisols*.

Cryosols are introduced at the highest level to identify a group of soils which occur under the unique environmental conditions of alternating thawing and freezing. These soils have permafrost within 100 cm of the soil surface and are saturated with water during the period of thaw. In addition, they show evidence of cryoturbation. *Durisols* comprise the soils in semi-arid environments which have an accumulation of secondary silica, either in the form of nodules, or as a massive, indurated layer. *Umbrisols* cover the soils which have either an umbric horizon, or have a mollic horizon and a base saturation of less than 50 per cent in some parts within the upper 125 cm of the soil surface. They are a logical counterpart of the *Chernozems, Kastanozems* and *Phaeozems*.

The *Plinthosols* bring together the *Plinthosols* of the Revised Legend and the soils which have a petroplinthic layer at shallow depth. In the Revised Legend the latter soils belong to the *Leptosols*. For the World Reference Base it was decided to exclude from the *Leptosols* soils with pedogenetic horizons such as indurated calcic or gypsic horizons or hardened plinthite. This necessitated the definition of a reference soil group which included these soils. Although it is realised that soils with shallow petroplinthic layers and soils having plinthite normally occupy different positions in the landscape, it was felt appropriate to group them together as they are genetically related.

Podzoluvisols are renamed *Albeluvisols*. The name *Podzoluvisols* suggests that in these soils both the processes of cheluviation (leading to *Podzols*) and subsurface accumulation of clay (resulting in *Luvisols*) take place, while in fact the dominant process consists of removal of clay and iron/manganese along preferential zones (pea faces, cracks) in the argic horizon. The name *Albeluvisols* is therefore thought to be more appropriate, expressing the presence of a bleached eluvial horizon ("*albic horizon*"), a clay-enriched horizon ("*argic horizon*") and the occurrence of "*albeluvic tonguing*".

WRB DIAGNOSTIC HORIZONS, PROPERTIES AND MATERIALS

Earlier it was agreed that the soil groups should be defined in terms of a specific combination of soil horizons, called '*reference horizons*' rather than '*diagnostic horizons*'. Reference horizons were intended to reflect genetic horizons which are widely recognized as occurring in soils. Unfortunately, the distinction between reference and diagnostic horizons created confusion and it was agreed to retain the FAO terminology of diagnostic horizons as well as the diagnostic properties. Additionally it appeared necessary to define diagnostic soil materials. This together resulted in a comprehensive list of WRB diagnostic horizons, properties and materials, defined in terms of morphological characteristics and/or analytical criteria. In line with the WRB objectives, attributes are described as much as possible to help field identification.

MODIFICATIONS TO DEFINITIONS OF FAO'S DIAGNOSTIC HORIZONS AND PROPERTIES

Of the 16 diagnostic horizons of the Revised Legend only *the fimic A horizon* has not been retained. It covers too wide a range of human-made surface layers and is replaced in the WRB by the hortic, plaggic and ferric horizons.

For the WRB, the definition of the *histic horizon* was broadened by reducing its minimum thickness to 10 cm and removing the maximum thickness. This is because of a second use of the definition. In the Revised Legend the histic H horizon is used to distinguish soils at second level to identify histic soil units; in the WRB it is used also at the highest level to define *Histosols*. It was agreed that *Histosols* over continuous hard rock should have a minimum thickness of 10 cm in order to avoid very thin organic layers over rock being classified as *Histosols*. The P_2O_5 content requirement for FAO's mollic and umbric A horizons has been deleted from the WRB definition of *mollic* and *umbric horizons*. This requirement cannot be considered diagnostic since thick, dark coloured, human-made horizons in, for instance, China, also have low amounts of phosphate. Other criteria have to be found to separate mollic and umbric horizons from anthropedogenic horizons.

A *chernic horizon* is defined as a special kind of mollic horizon. The present definition of the mollic horizon was felt to be too broad to reflect properly the unique characteristics of the deep, blackish, porous surface horizons which are so typical for *Chernozems*.

The definition of the *ochric horizon* is similar to the ochric A horizon. The colour requirement for the *albic horizon* have been slightly changed compared to FAO's albic E horizon, to suit albic horizons which show a considerable shift in chrome upon moistening. Such conditions are frequently found in soils of the southern hemisphere.

The *argic horizon* definition differs from that of the argic B horizon of the Revised Legend in that the percentage clay skins on both horizontal and vertical ped faces and in pores has been increased from one to five per cent. This is expected to provide a better correlation with the earlier requirement of at least one per cent oriented clay in thin sections.

Guidelines to recognize a lithological discontinuity, if not clear from the field observation, were added to the description of the argic horizon. It can be identified by the percentage of coarse sand, fine sand and silt, calculated on a clay-free basis (international particle size distribution or using the additional groupings of the United States Department of Agriculture (USDA) system or other), or by changes in the content of gravel and coarser fractions. A relative change of at least 20 per cent in any of the major particle size fractions is regarded as diagnostic for a lithological discontinuity. However, it should only be taken into account if it is located in the section of the solum where the clay increase occurs and if there is evidence that the overlying layer was coarser textured.

The adjustments made in the description of the argic horizon also apply to the *natric horizon*. The definition of FAO's cambic B horizon has been slightly amended by deleting the requirement '*....and has at least eight per cent clay*'. This requirement forces some soils, which have a well-developed structural-B horizon and silt loam or silt textures with a low clay content, as found, for instance, in fluvio-glacial deposits of the nordic countries, into the *Regosols* rather than in the *Cambisols*. Because there is also no need for this requirement to separate *Cambisols* from*Arenosols* it has not been used in the definition proposed for the WRB *cambic horizon*.

Major alterations are made in the definition of the *spodic horizon*. It has been brought into line with the recent modifications in soil taxonomy regarding the definition of spodic materials. Colour requirements were added, a limit of 0.5 or more in percentage oxalate extractable aluminium plus half that of iron is used, and a value for the optical density of oxalate extract (ODOE) of 0.25 or more is introduced. Moreover, the upper limit of spodic horizons has been set at 10 cm depth. The silt-clay ratio of 0.2 or less has been deleted from the definition of the *ferralic horizon*. This criterion was felt to be too strict; the silt particle size fraction has been increased. Other values have been proposed (silt-clay ratio of 0.7 or less; fine silt-clay ratio of 0.2 or less) but, as yet, no consensus has been reached.

Some alterations are made in the definitions of the *calcic* and *gypsic horizons*. For WRB purposes they are split into calcic/gypsic and*hypercalcic/hypergypsic horizons*. These latter horizons have a calcium carbonate equivalent and gypsum content of 50 and 60 per cent, respectively, but are not cemented. The definition for the *sulfuric horizon* remains the same as in the Revised Legend. In addition to these diagnostic horizons, 19 new ones are proposed. Some are adopted from FAO's diagnostic properties, others are newly formulated. Together they bring the total of diagnostic horizons recognized in the WRB to 34. The newly defined diagnostic horizons are the *andic, anthropedogenic, chernic, cryic, duric, ferric, folic, fragic, fulvic, glacic, melanic, nitic, petroduric, petroplinthic, plinthic, salic, takyric, vertic, vitric* and *yermic horizons*.

A combination of an *anthraquic horizon* at the surface with an underlying *hydragric horizon*, totalling together a thickness of at least 50 cm, defines certain*Anthrosols* which show evidence of alteration through wet-cultivation practices. It comprises a puddled layer, a plough pan and an illuvial subsurface horizon. This combination is characteristic for soils which have been used for long-term paddy rice cultivation.

Newly defined diagnostic properties and materials are *albeluvic tonguing, alic* and *aridic properties*, and *anthropogeomorphic, calcaric, fluvic, gypsiric, organic, sulfidic* and *tephric soil material*.

Gleyic and *stagnic properties* have been reformulated. Slight changes are made in FAO's definitions of *abrupt textural change and geric properites*, while the definitions of *permafrost* and soft powdery lime, renamed *secondary carbonates*, have been adopted without change.

In the description of the *gleyic* and *stagnic properties* the occurrence of 'gleyic' and 'stagnic colour patterns' is introduced. These terms apply to the specific distribution pattern of Fe/Mn (hydr)oxides caused by saturation with groundwater or stagnating surface water. A gleyic colour pattern has 'oximorphic' features on the outside of structural elements, along root channels and pores, or as a gradient upwards in the soil. A stagnic colour pattern on the other hand shows these features in the centre of peas or as a gradient downwards resulting from impedance of the water flow.

The slight changes in the descriptions of *abrupt textural change* and *geric properties* refer to a different depth in which tile change in texture must occur and another way of calculating the effective cation exchange capacity (ECEC)[1], respectively. Since the inception of the Legend of the Soil Map of the World, the number of lower level units used in the Legend or soil classification has continued to grow: from 106 in 1974 to 152 in the Revised Legend of the Soil Map of the World to 209 in the first draft of the *World Reference Base for Soil*

Resources. At the same time a serious effort was undertaken to expand this second level further with the introduction of third level units. Further proliferation of soil units and subunits in the World Reference Base might easily lead to a situation where it will become extremely difficult to recall and use all definitions within the main reference soil groups.

A further complication is that many soil unit names, and modifiers in the draft WRB, were inherited from the original FAO Legend and were defined depending on the grouping in which they occurred. For example, a "Dystric" soil unit may mean: "... having a base saturation of less than 75%" (in Dystric Vertisols), or "...having a base saturation of less than 50%", in different control sections (*e.g.* note the difference in control sections of Dystric Planosols and Dystric Cambisols). Another limitation inherent to the close link with the Legend of the Soil Map of the World is that, although often used as a soil classification system, the original purpose of the FAO system was to serve as a Legend for a specific map, which made certain simplifications necessary. For example, Calcic Gleysols included soil with a gypsic horizon. Similarly, Umbric Fluvisols grouped alluvial soils with an umbric horizon together with Fluvisols with a desaturated histic horizon. This resulted in loss of information due to the generalization required for the Legend.

Last but not least, it is thought that a clear-cut separation must be made between the double objectives of the World Reference Base, which on one hand should be able to serve as a soil reference system for geographers, agronomists and other users who are mainly interested in the highest level of generalization explained in non-technical terms, while on the other hand WRB must be a sophisticated tool for soil correlation able to accommodate a wide range of national soil classification systems.

PLANT CELL ORGANELLES

Plants are highly evolved eukaryotic organisms that comprise of membrane bound cell organelles. Even though plants and animals belong to eukaryotic groups, they differ in certain characteristic features. For example, a plant cell possesses a well-developed cell wall and large vacuoles, while an animal cell lacks such structural parts. In addition to these, a plant cell lacks centrioles and intermediate filaments, which are present in an animal cell.A typical plant cell is made up of cytoplasm and organelles. Scientific studies have been done regarding plant cell organelles and their functions. Each of the organelles of a plant cell has specific functions, without which the cell cannot operate properly.

LIST OF PLANT CELL ORGANELLES

When it comes to plant cell organelles, they are more or less similar to animal cells, except that the latter lacks chloroplast organelles, that are responsible for photosynthesis. Following is a list of organelles found in plant cell:

Nucleus

Nucleus (plural nuclei) is a highly specialized cell organelle, which stores the genetic component (chromosomes) of the particular cell. It serves as the main administrative centre of the cell, by coordinating the metabolic processes like cell growth, cell division and protein synthesis.

Plastids

Plastids are collective terms for organelles that carry pigments. In a plant cell, chloroplasts are the most prominent forms of plastids, which contain the green chlorophyll pigment. Because of these chloroplast plastids, a plant cell has the ability to undergo photosynthesis in the presence of sunlight and synthesize its own food.

Ribosomes

Ribosomes are plant cell organelles that comprise of proteins (40 per cent) and ribonucleic acid or RNA (60 per cent). They are important organelles responsible for the synthesis of proteins. Each ribosome consists of two parts, a larger subunit and a smaller subunit.

Mitochondria

Mitochondria (singular mitochondrion) are spherical to rod-shaped organelles present in the cytoplasm of the plant cell. They break down the complex carbohydrates and sugars into usable forms, for the plant. As mitochondria indirectly supply energy for the plant cell, they are also called as the powerhouse of the cell.

Golgi Body

Golgi body is also referred to as golgi complex or golgi apparatus. It plays a major role in transporting chemical substances in and out of the cell. After the endoplasmic reticulum synthesizes lipids and proteins, golgi body alters and prepares them for exporting outside the cell.

Endoplasmic Reticulum

Endoplasmic reticulum (ER) is the connecting link between the nucleus and cytoplasm of the plant cell. Basically, it is a network of interconnected and convoluted sacs that are located in the cytoplasm. Based on the presence or absence of ribosomes, ER can be of smooth or rough types. The former type lacks ribosomes, while the latter is covered with ribosomes. Overall, endoplasmic reticulum serves as a manufacturing, storing and transporting structure for glycogen, proteins, steroids and other compounds.

Vacuoles

Vacuoles are the storage organelles that help in regulating turgor pressure of the plant cell. In a plant cell, there can be more than one vacuole. However, the centrally located vacuole is larger than others, which stores all sorts of chemical compounds. Vacuoles also assist in intracellular digestion of complex molecules and excretion of waste products.

Peroxisomes

Peroxisomes are cytoplasmic organelles of the plant cell, which contains certain oxidative enzymes. These enzymes are used for the metabolic breakdown of fatty acids into simple sugar forms. Another important function of peroxisomes is to help chloroplasts in undergoing photorespiration process.Well! This was brief information regarding plant cell organelles, their structure and specific functions. As we have seen, coordination of the functions of plant cell organelles is crucial for carrying out the physiological and biochemical functionalities of the plant.

Cell Nucleus: Structure

The structure of a cell nucleus consists of nuclear membrane (nuclear envelope), nucleoplasm, nucleolus and chromosomes. Nucleoplasm, also known as karyoplasm, is the matrix present inside the nucleus. Let's discuss in brief about the several parts of a cell nucleus.

Nuclear Membrane

The nuclear membrane is a double-layered structure that encloses the contents of the nucleus. The outer layer of the nuclear membrane is connected to the endoplasmic reticulum. A fluid-filled space or perinuclear space is present between the two layers of a nuclear membrane. The nucleus communicates with the remaining of the cell or cytoplasm through several openings called nuclear pores. Nuclear pores are the sites for the exchange of large molecules (proteins and RNA) between the nucleus and cytoplasm.

Chromosomes

Chromosomes are present in the form of strings of DNA and histones (protein molecules) called chromatin. Chromatin is further classified into heterochromatin and euchromatin based on the function. The former type is a highly condensed trancriptionally inactive form, mostly present in adjacent to the nuclear membrane. Euchromatin is a delicate, less condensed organization of chromatin, which is found abundantly in a transcribing cell.

Nucleolus

The nucleolus is a dense, spherical-shaped structure present inside the nucleus. Some of the eukaryotic organisms have nucleus that contains up to four nucleoli. The nucleolus plays an indirect role in protein synthesis by producing ribosomes. Ribosomes are cell organelles made up of RNA and proteins; they are transported to the cytoplasm, which are then attached to the endoplasmic reticulum. Ribosomes are the protein-producing structures of a cell. Nucleolus disappears when a cell undergoes division and is reformed after the completion of cell-division.

Cell Nucleus: Functions

Speaking about the functions of a cell nucleus, it controls the hereditary characteristics of an organism and is responsible for the protein synthesis, cell division, growth and differentiation.

Here is a list of the functions carried out by a cell nucleus:

- Storage of hereditary material, the genes in the form of long and thin DNA (deoxyribonucleic acid) strands, referred to as chromatins.
- Storage of proteins and RNA (ribonucleic acid) in the nucleolus.
- Nucleus is a site for transcription in which messenger RNA (MRNA) are produced for the protein synthesis.
- Exchange of hereditary molecules (DNA and RNA) between the nucleus and rest of the cell.
- During the cell division, chromatins are arranged into chromosomes.
- Production of ribosomes (protein factories) in the nucleolus.
- Selective transportation of regulatory factors and energy molecules through nuclear pores.

As the nucleus regulates the integrity of genes and gene expression, it is also referred to as the control centre of a cell. Overall, the cell nucleus stores all the chromosomal DNA of an organism.The environment which we inhabit and live our daily lives, is an incorporation of several kinds of components.

These components, although may not share any resemblance, yet are in a mutual dependency with each other. These components not only refer to the living things, but other elements which are classified as the non-living things on the Earth. Both these elements are important in their own domain in the environment, as they work to maintain the delicate balance in the ecosystem. And this very balance helps to regulate all those things which makes life possible on the Earth. Ergo, with any of these things missing in this system, the biological organization will suffer nothing less than chaotic events. So let's proceed further to understand these types of components and what role do they play in the levels of organization of life.

LEVELS OF ORGANIZATION

In an ascending manner, the levels of biological organization are as follows:

Subatomic Particles

Any substance is composed mainly of three subatomic particles. They include the protons, neutrons and electrons. As you must be aware, protons refer to positively charged particles, while neutrons are ones which posses no charge. These two kinds reside inside the nucleus of the atom. Now, about the electrons, they posses negative charge and rotate around the nucleus, in different energy shells. Another subatomic particle is what is known as the photon - possessing zero mass and rest energy and traverses with the speed of light. It is actually defined as a quantum of electromagnetic energy.

Atoms

After the subatomic particles, now we come to the next point in the levels of organization, know as atoms. All objects are made up of matter in this universe. And the basic building blocks of these matter are known as atoms. These components consist of equal number of protons and neutrons. However, a single element might have two different atoms, wherein, there might be a difference in the number of neutrons. The centre of the atom is known as the nucleus, and it comprises of protons and neutrons. This structure of the atomic nucleus is orbited by negatively charged electrons.

Small Molecules

By saying small molecules, I am referring to those levels of organization, which comprises of amino acids, fatty acids, glucose, etc.

Macromolecules

Now these are the large molecules which are contained in a cell, and are required for carrying out the vital processes of life. Examples include carbohydrates, lipids, proteins, nucleic acids, etc.

Molecular Assemblies

A level higher than macromolecules, these assemblies are defined as sets which comprise of one or more molecular entity.

Organelle

Every cell has a specialized part which serve as organs. These are known as organelle. Examples are nucleus, mitochondria, etc.

Cells

This is one of the important levels of organization, wherein, cells are known to play vital roles. Every living creature on this Earth consists of cells, which are primarily, their structural and functional units. For instance, microorganisms like bacteria are known to be single-celled organisms. While, we humans and other organisms, are known to have as many as 100,000,000,000,000 cells!

Tissues

Tissue is nothing but a term that is used to refer to a group of cells. There are several functions in the body of living organisms. And different function is assigned to different tissues. For example, muscle tissue, nervous tissue, etc., are found in animals, while plants, have meristematic tissues and permanent tissues.

Organs

Again, when tissues are grouped together, they form what is known as an organ, which also has an important place in the levels of organization. As you are aware, heart, lungs, kidneys, stomach are organs. And these organs are enabled by the different tissues to carry out their specific duties.

Organ System

As you must have guessed already, this system is a formed by different kinds of organs. The system is assigned to perform higher levels of work in the body. This is more important in multicellular organisms like us and animals. A very simple example of an organ system is the circulatory system. It is a collaboration of heart, blood and blood vessels.

Organisms

Now when all the above levels of organization are taken in one picture, what is derived is known as organism like plants, animals, humans, bacterium, etc.

Population

It is here that we speak about a broader concept in the levels of organization of living things. A group of organisms, belonging to the same species may have the capability to inter-breed with each other. And when they inhabit a particular area, they are known as population. Its example can be the population of frogs in a pond.

Species

Species is a term that is used to mark the distinction between different kinds of organisms. For example, human species are different from animal species and species of plants.

Community

Taking it down to the last three of the levels of organization, we are speaking of community. It is nothing but a reference to a group of organism of different species, which populate a given area and stay in mutual interaction with each other.

Ecosystem

To define it in a layman language, interaction of a group of organism with the environment that they live in, is an ecosystem. For example, a pond ecosystem may consists of organisms such as fish, frogs, insects, etc., who depend on the water, mud, plants, etc., for survival.

Biosphere

Finally, we have reached the last block of the pyramid of the levels of organization, with what is known as biosphere. It encompasses all. Meaning, it consists of the regions which lie on, above and below the surface of the Earth, including the atmosphere where the living organisms thrive.

So there you are with a quick explanation on the different levels of organization, which we, along with several other components are living. Each level serves as a building block for the other, and this is what lays emphasis on the importance of all the components.

CELL CULTURE IN PLANT TISSUE

Plant tissue culture encompasses culturing of plant parts on an artificial medium. The plant parts can be a single cell, tissue or an organ. It is also referred to as micropropagation. Plant tissue culture was practically implemented for the first time by Haberlandt, a German scientist, in 1902. Later in 1934, Gautheret found successful results on in-vitro culture of plants. The basic key used in plant tissue culture is the totipotency of plant cells, meaning that each plant cell has the potential to regenerate into a complete plant. With this characteristic, plant tissue culture is used to produce genetically identical plants (clones) in the absence of fertilization, pollination or seeds.

Plant tissue culture encompasses culturing of plant parts on an artificial medium. The plant parts can be a single cell, tissue or an organ. It is also referred to as micropropagation. Plant tissue culture was practically implemented for the first time by Haberlandt, a German scientist, in 1902. Later in 1934, Gautheret found successful results on in-vitro culture of plants. The basic key used in plant tissue culture is the totipotency of plant cells, meaning that each plant cell has the potential to regenerate into a complete plant. With this characteristic, plant tissue culture is used to produce genetically identical plants (clones) in the absence of fertilization, pollination or seeds.

METHODS AND TECHNIQUES

Raupp et al. evaluated the relative yield, product quality and soil life after long-term (17 years) organic or mineral fertilization. They found that production yields varied: yields of potatoes and rye were lower and yields of spring wheat were similar in the organic system. However, their evaluation also showed that humus content and biological activity in the soil were greater, that products from the organic system had a better storage quality and that vegetables had lower nitrate contents.

Apart from the quality of the product from different production systems, the production potential of organic in comparison with conventional production systems is important. Research findings seem contradictory as production yields are found to be lower, equivalent, or higher. Simulating the production potential after long term differing management systems showed the following: Droogers et al. compared the production potential of two farming systems (biodynamic and conventional) by converting 'static' basic soil properties into a 'dynamic' assessment using simulation modelling.

Soil conditions on two farms - one having been managed biodynamically for 70 years - were investigated with morphological and physical methods. A simulation model including 30 years climatic data was used to predict water-limited potato yields, showing that the simulated yields were significantly higher on the biodynamic fields.

RHIZOSPHERE - INTERACTION WITH PLANT GROWTH

No published research was identified in New Zealand on this topic. The international work published emphasized the influence that soil flora and fauna have on plant growth in an organic system. Scullion et al., (1998), for example, described how management before conversion can have an impact on soil fungi, that in turn will influence plant growth under the organic system. Yeates et al.,described soil microbe and faunal diversity in Welsh soils, and while the soil types may differ, the message to enhance diversity is the same. This chapter also described indicators that could be used to define soil "health" in an organic system. Eason et al.and Cook et al.,both described how management can impact on soil microbe and fungi populations and how this in turn can influence plant growth.

TRACE ELEMENTS

Condron et al presented a comparison of soil quality under conventional and organic management in New Zealand. One area of focus in the paper was on trace elements in soils under organic and conventional pasture and their effects on animal health. Trace element impacts on animal health were also mentioned in some of the more generic animal health papers described earlier. Other than this work, there was very little information on trace elements in soils under organic pastures.

MICROBIAL ACTIVITIES AND BIOMASS

Synthetic water soluble fertilizers are known to be detrimental to soil microorganisms and therefore to influence biological and physical soil characteristics negatively in the long term. Mader et al.studied the effect different fertilization intensities on different crops have on soil microorganisms at different soil depths. They compared organic and conventional farming systems in Switzerland and reported that soil microbial biomass was significantly higher under the biodynamic system. Wood reported on studies in the UK, that showed an increase in earthworm populations with additions of manure; and comparisons of conventional and biodynamic farms in New Zealand showed that the biodynamically farmed soils had better structure, lower bulk density, higher organic matter content and respiration rates, and higher earthworm populations.

Pfiffner et al described a long-term trial studying the effect of different farming systems (biodynamic, organic and conventional) on ground beetles. The number of species was consistently higher (193%) in the biodynamic than in the conventional (100%) and organic (188%) plots. The number of species differed from 18-24 in biodynamic plots, compared to 19-22 in organic and 13-16 in conventional plots. In 1998 Pfiffner et al., (1998) reported on a long-term trial into the effect of different farming systems (biodynamic, organic and conventional) on earthworm populations. The earthworm biomass and density, and the number of juveniles and earthworm species were significantly higher in the biological than the conventional or organic plots.

Ryan reviewed studies with regard to the question 'Is an enhanced soil biological community a consistent feature of alternative agricultural systems?' Studies that examined four groups of soil organisms, comparing the soil biological community in conventional, organic and biodynamic management system were reviewed, and a case study of biodynamic and conventional dairy farms in Australia included.

Ryan's conclusion was that the enhanced soil community found on the biodynamic farms relative to conventional neighbours should not be considered as a definitive feature of alternative agricultural systems, but rather as an effect of the higher input of organic matter. Lytton-Hitchins et al.came to the same conclusion after comparing the physical and chemical properties of biodynamically and conventionally managed pastures in NE Victoria, Australia. The more favourable properties found on the biodynamically managed soils are attributed to decreased grazing pressure, longer intervals between irrigations, reduced tractor traffic and intermittent applications of compost and hornmanure preparations.

Ryan et al., undertook a glasshouse experiment with soil samples from 3 biodynamic (without conventional fertilizers for 17 years) and conventional (regular inputs of soluble P and N fertilizers) dairy pastures. Plant nutrient uptake was examined by assessing the response of white clover, perennial rye grass and indigenous VAM fungi to the addition of 4 levels of soluble P and N. Their findings indicated that the soils had not developed substantially different processes to enhance plant nutrient uptake. The experiments did not examine the response of plant nutrient uptake to non-water-soluble fertilizer, which might have produced different results.

PLANT CELL MODEL

Biotechnology is name given to the methods and techniques that involve the use of living organisms like bacteria, yeast, plant cells etc or their parts or products as tools (for example, genes and enzymes). They are used in a number of fields: food processing, agriculture, pharmaceutics, and medicine, among others. Plant tissue culture can be defined as culture of plant seeds, organs, explants, tissues, cells, or protoplasts on nutrient media under sterile conditions.

The science of plant tissue culture takes its roots from path breaking research in botany like discovery of cell followed by propounding of cell theory. In 1839, Schleiden and Schwann proposed that cell is the basic unit of organisms. They visualized that cell is capable of autonomy and therefore it should be possible for each cell if given an environment to regenerate into whole plant. Based on this premise, in 1902, a German physiologist, Gottlieb Haberlandt developed the concept of in vitro cell culture.

He isolated single fully differentiated individual plant cells from different plant species like palisade cells from leaves of Laminum purpureum, glandular hair of Pulmonaria and pith cells from petioles of Eicchornia crassiples etc and was first to culture them in Knop's salt solution enriched with glucose. In his cultures, cells increased in size, accumulated starch but failed to divide. Therefore, Haberlandt's prediction failed that the cultured plant cells could grow, divide and develop into embryo and then to whole plant.

This potential of a cell is known as totipotency, a term coined by Steward in 1968. Despite lack of success, Haberlandt made several predictions about the requirements in media in experimental conditions which could possibly induce cell division, proliferation and embryo induction. G Haberlandt is thus regarded as father of tissue culture. Taking cue from Haberlandt's failure, Hannig chose embryogenic tissue to culture. He excised nearly mature embryos from seeds of several species of crucifers and successfully grew them to maturity on mineral salts and sugar solution. In 1908, Simon regenerated callus, buds and roots from Poplar stem segments and established the basis for callus culture.

For about next 30 years, there was very little further progress in cell culture research. Within this period, an innovative approach to tissue culture using meristematic cells like root and stem tips was reported by Kolte and Robbins working independently.All these research attempts involving culture of isolated cells, root tips or stem tips ended in development of calluses. There were two objectives to be achieved before putting Haberlandt's prediction to fruition.

First, to make the callus obtained from the explants to proliferate endlessly and second to induce these regenerated calluses to undergo organogenesis and form whole plants. It was in 1930s, when progress in plant tissue culture accelerated rapidly owing to an important discovery that vitamin B and natural auxins were necessary for the growth of isolated tissues containing meristems.

This breakthrough came from White who reported that not only could cultured tomato root tips grow but could be repeatedly subcultured to fresh medium of inorganic salts supplemented with yeast extract. He later replaced YE by vitamin B namely pyridoxine, thiamine and proved their growth promoting effect.

DESCRIPTION

Plant cells are the cells found in plants comprising sub cellular organelles. Following is a plant cell model for kids that will give your child a visual of the plant cell and a better understanding of its various parts.

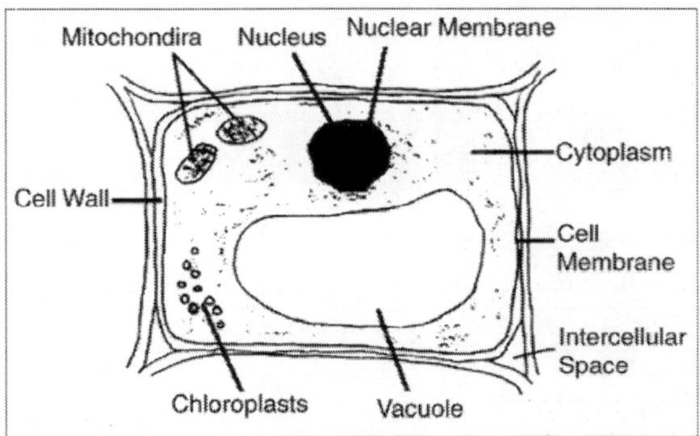

The cells are the basic units of life that work together to perform life sustaining functions in both animal and plant worlds.

Plant cells are eukaryotic cells having thick and rigid cell walls. Eukaryotic cells are cells that contain complex structures enclosed within membranes called nucleus or nucleus envelop, within which the genetic material is present.

These cells are present in almost all living organisms including animals, plants and fungi.

PLANT CELL MODEL AND PARTS

Plant cells differ from animal cells in that they have three different structures known as cell wall, vacuoles and plastids. However, they lack centrioles and intermediate filaments, which are present in animal cells. Read more on plant cell structure and parts.

Cell Membrane

It permits the waste material to exit the cell and regulates the movement of materials in and out of the cell. It also acts as a barrier between the inside of the cell and the outside, so that the chemical conditions on both sides can be different.

Cell Wall

It provides rigid structural support to the cell from which structures, like, leaves and stems are produced. It allows the circulation and distribution of water, minerals and nutrients in and outside the cell. The wall also controls the presence of pathogen microbes and development of tissues in the cell.

Golgi Body

It stores, packages, and circulates the lipids and proteins made in the endoplasmic reticulum. It stores the proteins into packages called vesicles.

Rough Endoplasmic Reticulum

It synthesizes and exports proteins and glycoproteins throughout the cell.

Lysosomes

These organelles digest food by breaking down larger molecules into smaller ones, using special proteins.

Cytoplasm

It helps in maintaining the cell shape. Along with this, it plays an important role in the internal movement of cell organelles, cell mobility and muscle fibre contraction.

It also distributes oxygen and nutrients to different parts of the cell.

Nucleolus

It is the most prominent structure in the nucleus wherein ribosomes are made.

Vacuole

The plant cell contains a large, single vacuole (an enclosed compartment), which is used to store water, compounds and minerals that help in plant growth.

Ribosomes

These are packets of RNA (Ribonucleic Acid) and are called as protein builders or synthesizers of the cell.

Chloroplasts

They are the food producers of the cell, containing chlorophyll, the green pigment essential for photosynthesis. Their main function is to generate sugars and starches.

Nucleus

It's the most important part of the cell, comprising chromosomes, *i.e.* structures made up of genetic information that helps in cell growth and reproduction. Read more on cell nucleus: structure and functions.

Nuclear Envelope

It's an enclosure that surrounds nucleus and its contents. Unlike cell membrane, which has pores and spaces for RNA and proteins to pass through, it keeps the chromatin and nucleolus inside the nucleus.

Smooth Endoplasmic Reticulum

It's main function is to package proteins for transport, synthesize membrane phosolipids, and secrete calcium. It also performs transformation of bile pigments, glycogenolysis (the breakdown of glycogen), and detoxification of different drugs and chemical agents.

Mitochondria

It provides energy to the cell by combining sugar molecules with oxygen to generate carbon dioxide and water.

Amylosplast

It's an organelle present in some plant cells that stores starch.

Druse Crystal

It's a granular type of crystal found in plant vacuoles. It is composed of calcium oxalate and is considered to deter herbivory.

Centrosome

Also known as Microtubule Organizing Centre, it's a region in the cell where microtubules are produced that perform a variety of functions ranging from transportation to structural support. Though both plant and animal cell centrosomes play similar roles during cell division, plant cell centrosomes are simpler and do not contain centrioles.

Peroxisomes

These are membrane bound packets of oxidative enzymes that convert fatty acids into sugar and assist chloroplasts in photo-respiration.

Golgi Vesicles

These vesicles transfer proteins and lipids to different parts of the cells.

PLANT CELL MODEL IDEAS

Following are some simple instructions or plant cell model ideas, to construct a model of a plant cell:

- Enumerate and collect all possible materials required for the model and look out for the different cell parts in a biology textbook.
- Use a heavy sturdy wooden base to construct the base of the model so that it can hold the weight and facilitate presentation of the model.
- Construct the cell nucleus and ensure that it's round in shape. Paint the nucleus in one colour so that it can be identified as a distinct cell part.

- With the help of a dowel rod or thick, straight stick, connect the nucleus to the base of the model. For this, drill a hole into the wooden base, stick the dowel rod into the hole and connect the nucleus to the rod.
- The cell wall and cell membrane of the plant cell model should be rectangular in shape. The outer wall is the cell wall, which should be made of a strong material like hard plastic or wood, and the cell membrane can be made of cellophane or thin plastic. Now attach both the walls to the nucleus and the model base.
- Add various other key components of the plant cell to the model. Like, for chlorophyll, use a piece of green fabric stuffed with cotton. Sew it and attach it to the model.
- Finally, prepare a report on the various plant cell parts and their corresponding functions.

TISSUE CULTURE OF LILACS

Traditionally lilacs have been grown from seed, suckers or by grafting superior clones on to compatible rootstocks. Lilac seedlings of several species, privet stem cuttings as well as ash seedlings/root cuttings have all been utilized as rootstocks.In recent times there has been a move away from using rootstocks for various reasons, and nurserymen have been trying to produce cultivars on their own roots. While cuttings taken at the right time are often successful for many species, this is not always a sure recipe for thevulgaris clones.For more than two decades some of the leading nurseries in the US and Europe have been using tissue culture for propagating lilacs on their own roots. While the main reason for resorting to micropropagation is to produce large numbers of selected clones within a short period, we have discovered it to be no less effective in producing small numbers ofvulgaris cultivars, which are recalcitrant to initiating roots by conventional means.

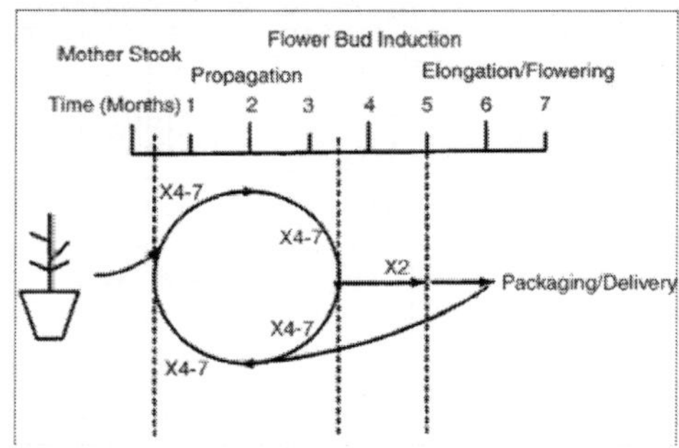

INVITRO FLOWERING OF ROSE

The present invention is directed to compositions and methods of micropropagation of a rose plant and in vitro flowering of the rose plant. Young shoots are induced to produce buds in an enclosed vessel containing a first culturing medium comprising benzyladenine, an auxin and 2% sucrose as carbon source.

Buds are excised and cultured on a medium for propagation and multiplication of plantlets. Plantlets are then transferred to a medium comprising thidiazuron, an auxin and myo-inositol for induction of flower buds, followed by culture on a medium comprising benzyladenine and an auxin for plantlet elongation and finally culture in the fifth medium without phytohormone for flowering. Alternately, after propagation and multiplication, plantlets are transferred to a medium comprising zeatin, auxin and myo-inositol for induction of flower buds followed by culture on a medium without phytohormone for elongation and flowering.

ORGANOGENESIS

Meristematic tissues are responsible for the division of new cells... they are zones of actively dividing cells.

Before we proceed, we should first recognize that growth in plants includes two stages:

1. First the production of new cells and,
2. Secondly the expansion of these cells via uptake of water by the vacuole.

Cell division occurs solely in meristematic regions, while expansion may occur anywhere. Thus in a single plant there are zones of young dividing cells, maturing cells, and mature cells. We will go into more detail on meristematic regions when we go over the stem and root individually, but for now an introduction is sufficient

Let us recognize that there are 3 meristematic regions in the plant:

I. Apical meristems are located at the apices or tips - at root and shoot tips and are directly involved in their elongation

They create derivatives which form primary growth:

- The protoderm which forms the outer dermal layer of tissues,
- The ground meristem which forms the cortical cells and
- The procambium which forms the vascular tissue.

In shoots, plants protect their meristems with young leaves and by forming dormant reserve meristems (*i.e.*, buds), that protect their apical meristems with a root cap

Ferns, Gymnosperms and Angiosperms are all terrestrial plants. Tissues therefore arise as a result of a cell or a group of cells dividing, thus giving rise to a large number of cells. In flowering plants these cells undergo changes that change the structure and function of the cell; differentiation has take place. As a result of differentiation, a cell or group of cells can have a specialized function; *i.e.* specialization takes place. Thus, a tissue is a group of cells that have differentiated to perform a specific function.

If a group of cells are more or the less the same, then it is known as a Simple tissue *e.g.* epidermis, parenchyma, collenchyma and sclerenchyma. If, however, the tissue is composed of a number of different cells, then it is known as a Complex tissue such as xylem and phloem.

DIAGRAMMATIC REPRESENTATION OF A PLANT BODY

Ground Tissue is simple non-meristematic tissue (non-dividing tissue) made up 3 cell types:paranenchyma, collenchyma and sclerenchyma.This tissue generally forms either the pith, cortex or bulk of leaf (mesophyll).

Parenchyma Cells

- Mostabundant cells in plants;
- Spherical cells which flatten at point of contact;
- Alive at maturity;pliable, primary cell walls;
- Large vacuoles for storage of starch, fats, and tannins (denature proteins);
- Primary sites of the metabolic functions such as photosynthesis, respiration, and protein synthesis; they are "ready reserves" from which a plant makes specialized cells to meet its changing needs.

SPECIALIZED PARENCHYMA

- *Chlorenchyma*: Photosynthetic cells; have high density of chloroplasts
- *Aerenchyma*: prominent intercellular spaces that improve gas exchange capacity of the tissue; provide maximum support with a minimum metabolic requirement
- *Transfer cells*: specialized for short distance transfer of solutes between cells; have secondary cell walls; they are inner extensions of wall that increase surface area.

Transfer cells occur in areas of high solute transport, such as secretory tissues, which release substances that are produced within the protoplasm and are moved outside, *i.e.*, nectar cells, mucilage in sundews, and resins.

Collenchyma Cells- Colla = Glue, Glisten

- Living protoplasm;
- Unevenly thickened primary cell walls; elongate cells;

- Longer than wide; just beneath epidermis; function to support growing organs, grass, floral parts, and border veins; their nonlignified cell walls can stretch

Sclerenchyma Cells- (skleos = hard)

- Most are dead at maturity;
- Rigid, thick, lignified, nonstretchable secondary cell walls. there are 2 types of sclerenchyma cells:

Sclereids or stone cells we saw in lab- short; variable shape; form hard layers such as the shells of nuts and seed coats; produce the gritty texture of pears. Fibres- long, slender; occur in strands or bundles; tiny cavity or lumen; the different hardnesses of fibres are used to make coarse rope, linen cloth, etc.

DIRECT SHOOT ORGANOGENESIS FROM LEAF EXPLANTS OF CHRYSANTHEMUM

Shoot regeneration from leaf and stem explants of 19 Chrysanthemum cultivars has been studied. Adventitious shoots could be regenerated from leaf and stem explants of 16 cultivars. Most cultivars showed the best shoot regeneration on Murashige and Skoog medium supplemented with 8.9 μM BAP and 2.7 μM NAA (0.5 mg/l). The number of regenerated shoots per explant was more than 10 in some of the cultivars studied. It has not been posssible to regenerate shoots from leaf explants of cultivars 170 and 260. The cultivars 140 and 8010 produced the highest number of shoots. Regenerated shoots can easily be rooted on half strength Murashige and Skoog medium supplemented with 0.98 μM IBA (0.2mg/l). After potting in soil and transfer to the glasshouse normal plants develop. Apparently no somaclonal variation has been observed among the regenerants.

Biolistic particle bombardment is used to transfer desirable genes into difficult to transform plants, but research is lacking on factors that can affect transformation success. This study was conducted to determine if manipulating culture conditions forRhododendron 'PJM Hybrid' and 'Iridon' chrysanthemum leaf explants could alter transient gene expression of e-glucuronidase (GUS) after bombardment. Both rhododendron and chrysanthemum leaf explants were bombarded with 1.0 μm gold particles coated with 2 μg of linear plasmid DNA, containing the GUS gene. Different explant treatments before bombardment were tested, including time in culture on regeneration medium, growing in the presence or absence of 0, 0.09, 0.2, 0.4, or 0.6 M sucrose, and growing in the dark for 0, 3, or 6 days. Transient GUS expression was histochemically quantified 2 days after bombardment as the percentages of GUS positive explants and the number of blue spots per GUS positive explant. Transient GUS expression was highest when 'PJM Hybrid' explants were incubated on regeneration medium 9 to 12 days before bombardment, whereas transient GUS expression by 'Iridon' explants was relatively high (>42%) for leaves incubated 3 to 18 days before bombardment. Exposure of 'PJM Hybrid' explants to media containing different concentrations of sucrose decreased transient GUS expression by 71% compared to those grown on sucrose-free medium. Alternatively, 27% more 'Iridon' explants transiently expressed GUS compared to those on sucrose-free medium, and the mean number of spots per GUS positive explant increased from 5 to 13. Transient GUS expression increased almost three fold when 'PJM Hybrid' explants were placed in the dark for six days compared those in the light. In contrast, the dark treatment had inconsistent effects on transient GUS expression by 'Iridon' explants. Manipulating explant culture conditions before biolistic bombardment influenced transient gene expression and illustrated that genotype-specific differences affect transformation potential.

PRODUCTION OF DISEASE-FREE PLANTS (VIRUS)

Tissue culture technology has contributed in a plant that is free from viruses. In plants that have been infected with the virus, the cells in the bud tip (meristem) is an area that is not infected with the virus. In this way the meristem will mengkulturkan obtained virus-free plants.

GENETIC TRANSFORMATION

Tissue culture techniques have become an important part in helping the success of plant genetic engineering (gene transfer). For example, bacterial gene transfer (such as cry genes from Bacillus thuringiensis) into the plant cells will be expressed after transgeniknya achieved plant regeneration.

The production of secondary metabolites, compounds: Plant cell culture can also be used to produce biochemical compounds (secondary metabolites) such as alkaloids, terpenoids, phenyl etc. propanoid. This technology is now available in industrial scale. For example, the commercial production of compounds "shikonin" from Lithospermum erythrorhizon cell culture. Plant cells can be grown in isolation from intact plants in tissue culture systems. The cells have the characteristics of callus cells, rather than other plant cell types. These are the cells that appear on cut surfaces when a plant is wounded and which gradually cover and seal the damaged area.

Pieces of plant tissue will slowly divide and grow into a colourless mass of cells if they are kept in special conditions.

These are:

- Initiated from the most appropriate plant tissue for the particular plant variety
- Presence of a high concentration of auxin and cytokinin growth regulators in the growth media

- A growth medium containing organic and inorganic compounds to sustain the cells
- Aseptic conditions during culture to exclude competition from microorganisms

The plant cells can grow on a solid surface as friable, pale-brown lumps (called callus), or as individual or small clusters of cells in a liquid medium called a suspension culture. These cells can be maintained indefinitely provided they are sub-cultured regularly into fresh growth medium.

Tissue culture cells generally lack the distinctive features of most plant cells. They have a small vacuole, lack chloroplasts and photosynthetic pathways and the structural or chemical features that distinguish so many cell types within the intact plant are absent. They are most similar to the undifferentiated cells found in meristematic regions which become fated to develop into each cell type as the plant grows. Tissue cultured cells can also be induced to re-differentiate into whole plants by alterations to the growth media.

Plant tissue cultures can be initiated from almost any part of a plant. The physiological state of the plant does have an influence on its response to attempts to initiate tissue culture. The parent plant must be healthy and free from obvious signs of disease or decay. The source, termed explant, may be dictated by the reason for carrying out the tissue culture. Younger tissue contains a higher proportion of actively dividing cells and is more responsive to a callus initiation programme.

The plants themselves must be actively growing, and not about to enter a period of dormancy. The exact conditions required to initiate and sustain plant cells in culture, or to regenerate intact plants from cultured cells, are different for each plant species. Each variety of a species will often have a particular set of cultural requirements. Despite all the knowledge that has been obtained about plant tissue culture during the twentieth century, these conditions have to be identified for each variety through experimentation.

Table. Secondary Metabolites Produced in HighLevels by Plant Cell Cultures

Compound	Plant Species	Yields (% dry wt)		Culture
	Culture	**Plant**	**Type***	
Shikonin	Lithospermum	20	1.5	s
erythrorhizon				
Ginsenoside	Panax ginseng	27	4.5	c
Anthraquinones	Morinda citrifolia	18	0.3	s
Ajmalicine	Catharanthus roseus	1.0	0.3	s
Rosmarinic acid	Coleus blumeii	15	3	s
Ubiquinone-10	Nicotiana tabacum	0.036	0.003	s
Diosgenin	Dioscorea deltoides	2	2	s
Benzylisoquinoline	Coptis japonica	11	5 - 10	s
	Alkaloids			
Berberine	Thalictrum minor	10	0.01	s
Berberine	Coptis japonica	10	2 - 4	s
Anthraquinones	Galium verum	5.4	1.2	s
Anthraquinones	Galium aparine	3.8	0.2	s

Nicotine	Nicotiana tabacum	3.4	2.0	c
Bisoclaurine	Stephania cepharantha	2.3	0.8	s
Tripdiolide	Tripteryqium wilfordii	0.05	0.001	s

* s = suspension; c = callus

As seen in many reviews, studies on the production of plant metabolites by callus and cell suspension cultures have been carried out on an increasing scale since the end of the 1950's. The large scale cultivation of tobacco and a variety of vegetable cells was examined from the late 1950's to early 1960's by Tulecke and Nickell at Pfizer Inc., Mandels *et al.* at the Natick Laboratories in the U.S. Army, Street *et al.* at the University of Leicester and Martin *et al.* at the National Research Council of Canada. Their results stimulated more recent studies on the industrial application of plant cell culture in many countries.

Since Japan has a highly developed fermentation technology, many industrial companies, in collaboration with some university groups, have tried to apply this technology for the commercial production of useful compounds. The Japan Tobacco Inc.'s interest involved around mass-production of tobacco cells as raw materials of cigarettes; the company established 20 kL fermentors which were the largest for plant cells in 1970's. Meiji Seika in Japan also elucidated the fundamentals of production of Panax ginseng cells in large volumes.

Researchers reported that cultured ginseng cells stimulated physiological activities in animals in a similar fashion as elicited by native ginseng roots. The work was followed by Nitto Denko Co. which has been manufacturing cell mass of ginseng commercially. The cells are used as health foods in Japan. Researchers of Kyowa Hakko conducted extensive pharmacological screening of numerous cell cultures and found various novel products of great interest, including plasmin inhibitory proteins in Scopolia japonica cells and plant virus inhibitors in cultured cells of Phytolacca americana and other species. The virus inhibitors of P. americana are now being studied by many research groups in the world because of their activity against AIDS and other animal viruses. Other firms such as Ajinomoto and Nippon Shin-yaku also made efforts to increase the level of accumulation of alkaloids, steroids and other secondary products in cultured cells.

Groups in Germany outlined very interesting approaches to industrial application in the meeting held in 1976 at Munich, and their excellent results encouraged researchers in other countries. For example, Zenk and his colleagues presented an impressive paper in which they successfully selected high-alkaloid producing cell lines of Catharanthus roseus using a method similar to the microbial mono-colony isolation technique. A number of laboratories in industries and universities followed Zenk's approach, and in fact, some researchers could increase significantly the level of secondary products such as ubiquinone-10, biotin, and various plant pigments produced by cell cultures. In 1982, the 5th International Congress of Plant Tissue and Cell Cultures was held in Japan and about papers presented there related to production of secondary metabolites in cultured cells and several papers seemed to be commercially promising such as production of shikonin by Fujita *et al.*) of Mitsui Petrochemical and that of several antitumor compounds by Misawa *et al.*

At subsequent International Congress of Plant Tissue and Cell Cultures in Minneapolis in 1986 and that in Amsterdam in 1990 as well as other meetings such as the Meeting of Primary and Secondary Metabolism of Plant Cell Cultures held in Giessen, Germany and the 4th and International Congress on Phytotherapy held in Munich, Germany in September, 1992, many compounds were shown to be accumulated by plant cell cultures and many different strategies were presented to increase their productivity. Means for production of those compounds include not only de novo synthesis but also biotransformation processes. A biotransformation process to produce e-methyl digoxin using Digitalis lanata cells studied by Reinhard and Alfermann in Germany was evaluated by Boehringer Mannheim Co. using 4 kL bioreactors although it has not yet been commercialized. A combination of a plant cell culture process and a simple chemical coupling reaction was invented by a Canadian company, Allelix, to manufacture vinblastine as a commercially feasible process. The technology is now being studied by Mitsui Petrochemical in Japan forcommercialization.

In spite of remarkable advances in plant cell culture technology, the production cost of metabolites is still high, as estimated by Zenk and by Goldstein. The former calculated that the cost would be U.S. $500 per kg of isolated product when it was accumulated in 1 g/L within a period of 15 days in 100 m batch culture. It is true that the cost has decreased since his estimation in 1974, and the producing ability of 1 g/L for 15 days is achievable in the case of some compounds such as rosmarinic acid, but it is still too expensive to produce food stuffs, food additives or pharmaceuticals which can be more easily produced by alternative ways such as chemical synthesis, extraction from plants or by microbial fermentation.

Zenk stated that industrial plant cell culture techniques would be introduced only if the plant product under consideration is produced at a price equal to or preferably lower than the field-produced product. The only factor determining the industrial realization of plant cell culture is the price by which a given product can be produced.

TRANSGENIC PLANTS WITH BENEFICIAL TRAITS

During the last decades, a tremendous progress has been made in the development of transgenic plants using the various techniques of genetic engineering. The plants, in which a functional foreign gene has been incorporated by any biotechnological methods that generally are not present in the plant, are called transgenic plants. As per estimates recorded in 2002, transgenic crops are cultivated world-wide on about 148 million acres (587 million hectares) land by about 5.5 million farmers. Transgenic plants have many beneficial traits like insect resistance, herbicide tolerance, delayed fruit ripening, improved oil quality, weed control etc.

STRESS TOLERANCE

Biotechnology strategies are being developed to overcome problems caused due to biotic stresses (viral, bacterial infections, pests and weeds) and abiotic stresses (physical actors such as temperature, humidity, salinity etc).

Abiotic Stress Tolerance

The plants show their abiotic stress response reactions by the production of stress related osmolytes like sugars (*e.g.* trehalose and fructans), sugar alcohols (*e.g.* mannitol), amino acids (*e.g.* proline, glycine, betaine) and certain proteins (*e.g.* antifreeze proteins). Transgenic plants have been produced which over express the genes for one or more of the above mentioned compounds. Such plants show increased tolerance to environmental stresses. Resistance to abiotic stresses includes stress induced by herbicides, temperature (heat, chilling, freezing), drought, salinity, ozone and intense light. These environmental stresses result in the destruction, deterioration of crop plants which leads to low crop productivity. Several strategies have been used and developed to build ressitance in the plants against these stresses.

HERBICIDE TOLERANCE

Weeds are unwanted plants which decrease the crop yields and by competing with crop plants for light, water and nutrients. Several biotechnological strategies for weed control are being used *e.g.* the over-production of herbicide target enzyme (usually in the chloroplast) in the plant which makes the plant insensitive to the herbicide. This is done by the introduction of a modified gene that encodes for a resistant form of the enzyme targeted by the herbicide in weeds and crop plants. Roundup Ready crop plants tolerant to herbicide-Roundup, is already being used commercially.

The biological manipulations using genetic engineering to develop herbicide resistant plants are:

- Over-expression of the target protein by integrating multiple copies of the gene or by using a strong promoter.,
- Enhancing the plant detoxification system which helps in reducing the effect of herbicide.,
- Detoxifying the herbicide by using a foreign gene., and
- Modification of the target protein by mutation.

Some of the examples are:

Glyphosate resistance - Glyphosate is a glycine derivative and is a herbicide which is found to be effective against the 76 of the world's worst 78 weeds. It kills the plant by being the competitive inhibitor of the enzyme 5-enoyl-pyruvylshikimate 3- phosphate synthase (EPSPS) in the shikimic acid pathway. Due to it's structural similarity with the substrate phosphoenol pyruvate, glyphosate binds more tightly with EPSPS and thus blocks the shikimic acid pathway.

Certain strategies were used to provide glyphosate resistance to plants:

- It was found that EPSPS gene was overexpressed in Petunia due to gene amplification. EPSPS gene was isolated from Petunia and introduced in to the other plants. These plants could tolerate glyphosate at a dose of 2- 4 times higher than that required to kill wild type plants.
  - *By using mutant EPSPS genes*: A single base substitution from C to T resulted in the change of an amino acid from proline to serine in EPSPS. The modified enzyme cannot bind to glyphosate and thus provides resistance.
  - The detoxification of glyphosate by introducing the gene (isolated from soil organism- Ochrobactrum anthropi) encoding for glyphosate oxidase into crop plants. The enzyme glyphosate oxidase converts glyphosate to glyoxylate and aminomethylphosponic acid. The transgenic plants exhibited very good glyphosate ressitance in the field.

Another Example is of Phosphinothricin Resistance

Phosphinothricin is a broad spectrum herbicide and is effective against broad-leafed weeds. It acts as a competitive inhibitor of the enzyme glutamine synthase which results in the inhibition of the enzyme glutamine synthase and accumulation of ammonia and finally the death of the plant. The disturbace in the glutamine synthesis also inhibits the photosynthetic activity.

The enzyme phosphinothricin acetyl transferase (which was first observed in Streptomyces sp in natural detoxifying mechanism against phosphinothricin) acetylates phosphinothricin, and thus inactivates the herbicide. The gene encoding for phosphinothricin acetyl transferase (bar gene) was introduced in transgenic maize and oil seed rape to provide resistance against phosphinothricin.

OTHER ABIOTIC STRESSES

The abiotic stresses due to temperature, drought, and salinity are collectively also known as water deficit stresses. The plants produce osmolytes or osmoprotectants to overcome the osmotic stress. The attempts are on to use genetic engineering strategies to increase the production of osmoprotectants in the plants.

The biosynthetic pathways for the production of many osmoprotectants have been established and genes coding the key enzymes have been isolated. *E.g.* Glycine betaine is a cellular osmolyte which is produced by the participation of a number of key enzymes like choline dehydrogenase, choline monooxygenase etc. The choline oxidase gene from Arthrobacter sp. was used to produce transgenic rice with high levels of glycine betaine giving tolerance against water deficit stress.

Scientists also developed cold-tolerant genes (around 20) in Arabidopsis when this plant was gradually exposed to slowly declining temperature. By introducing the coordinating gene (it encodes a protein which acts as transcription factor for regulating the expression of cold tolerant genes), expression of cold tolerant genes was triggered giving protection to the plants against the cold temperatures.

INSECT RESISTANCE

A variety of insects, mites and nematodes significantly reduce the yield and quality of the crop plants. The conventional method is to use synthetic pesticides, which also have severe effects on human health and environment. The transgenic technology uses an innovative and eco-friendly method to improve pest control management.About 40 genes obtained from microorganisms of higher plants and animals have been used to provide insect resistance in crop plantsThe first genes available for genetic engineering of crop plants for pest resistance were Cry genes (popularly known as Bt genes) from a bacterium Bacillus thuringiensis. These are specific to particular group of insect pests, and are not harmful to other useful insects like butter flies and silk worms. Transgenic crops with Bt genes (*e.g.* cotton, rice, maize, potato, tomato, brinjal, cauliflower, cabbage, etc.) have been developed.

This has proved to be an effective way of controlling the insect pests and has reduced the pesticide use. The most notable example is Bt cotton (which contains CrylAc gene) that is resistant to a notorious insect pest Bollworm (Helicoperpa armigera).. There are certain other insect resistant genes from other microorganisms which have been used for this purpose. Isopentenyl transferase gene from Agrobacterium tumefaciens has been introduced into tobacco and tomato. The transenic plants with this transgene were found to reduce the leaf consumption by tobacco hornworm and decrease the survival of peach potato aphid.

Certain genes from higher plants were also found to result in the synthesis of products possessing insecticidal activity. One of the examples is the Cowpea trypsin inhibitor gene (CpTi) which was introduced into tobacco, potato, and oilseed rape for develping transgenic plants. Earlier it was observed that the wild species of cowpea plants growing in Africa were resistant to attack by a wide range of insects. It was observed that the insecticidal protein was a trypsin inhibitor that was capable of destroying insects belonging to the orders Lepidoptera, Orthaptera etc. Cowpea trypsin inhibitor (CpTi) has no effect on mammalian trypsin, hence it is non-toxic to mammals.

VIRUS RESISTANCE

There are several strategies for engineering plants for viral resistance, and these utilizes the genes from virus itself (*e.g.* the viral coat protein gene). The virus-derived resistance has given promising results in a number of crop plants such as tobacco, tomato, potato, alfalfa, and papaya. The induction of virus resistance is done by employing virus-encoded genes-virus coat proteins, movement proteins, transmission proteins, satellite RNa, antisense RNAs, and ribozymes. The virus coat protein-mediated approach is the most successful one to provide virus resistance to plants. It was in 1986, transgenic tobacco plants expressing tobacco mosaic virus (TMV) coat protein gene were first developed. These plants exhibited high levels of resistance to TMV.

The transgenic plant providing coat protein-mediated resistance to virus are rice, potato, peanut, sugar beet, alfalfa etc. The viruses that have been used include alfalfa mosaic virus (AIMV), cucumber mosaic virus (CMV), potato virus X (PVX), potato virus Y (PVY) etc.

RESISTANCE AGAINST FUNGAL AND BACTERIAL INFECTIONS

As a defence strategy against the invading pathogens (fungi and bacteria) the plants accumulate low molecular weight proteins which are collectively known as pathogenesis-related (PR) proteins. Several transgenic crop plants with increased resistance to fungal pathogens are being raised with genes coding for the different compounds. One of the examples is the Glucanase enzyme that degrades the cell wall of many fungi. The most widely used glucanase is beta-1,4-glucanase.

The gene encoding for beta-1,4 glucanase has been isolated from barley, introduced, and expressed in transgenic tobacco plants. This gene provided good protection against soil-borne fungal pathogen Rhizoctonia solani. Lysozyme degrades chitin and peptidoglycan of cell wall, and in this way fungal infection can be reduced. Transgenic potato plants with lysozyme gene providing resistance to Eswinia carotovora have been developed.

DELAYED FRUIT RIPENING

The gas hormone, ethylene regulates the ripening of fruits, therefore, ripening can be slowed down by blocking or reducing ethylene production. This can be achieved by introducing ethylene forming gene(s) in a way that will suppress its own expression in the crop plant. Such fruits ripen very slowly (however, they can be ripen by ethylene application) and this helps in exporting the fruits to longer distances without spoilage due to longer-shelf life.

The most common example is the 'Flavr Savr' transgenic tomatoes, which were commercialized in U.S.A in 1994. The main strategy used was the antisense RNA approach. In the normal tomato plant, the PG gene (for the enzyme polygalacturonase) encodes a normal mRNA that produces the enzyme polygalacturonase which is involved in the fruit ripening.

The complimentary DNA of PG encodes for antisense mRNA, which is complimentary to normal (sense) mRNA. The hybridization between the sense and antisnse mRNAs renders the sense mRNA ineffective. Consequently, polygalacturonase is not produced causing delay in the fruit ripening. Similarly strategies have been developed to block the ethylene biosynthesis thereby reducing the fruit ripening. *E.g.* transgenic plants with antisense gene of ACC oxidase (an enzyme involved in the biosynthetic process of ethylene) have been developed. In these plants, production of ethylene was reduced by about 97% with a significant delay in the fruit ripening.

The bacterial gene encoding ACC deaminase (an enzyme that acts on ACC and removes amino group) has been transferred and expressed in tomato plants which showed 90% inhibition in the ethylene biosynthesis.

MALE STERILITY

The plants may inherit male sterility either from the nucleus or cytoplasm. It is possible to introduce male sterility through genetic manipulations while the female plants maintain fertility. In tobacco plants, these are created by introducing a gene coding for an enzyme (barnase, which is a RNA hydrolyzing enzyme) that inhibits pollen formation.

This gene is expressed specifically in the tapetal cells of anther using tapetal specific promoter TA29 to restrict its activity only to the cells involved in pollen production. The restoration of male fertility is done by introducing another gene barstar that suppresses the activity of barnase at the onset of the breeding season. By using this approach, transgenic plants of tobacco, cauliflower, cotton, tomato, corn, lettuce etc. with male sterility have been developed.

PLANT GERMPLASM CONSERVATION

Germplasm refers to the sum total of all the genes present in a crop and its related species.The conservation of germplasm involves the preservation of the genetic diversity of a particular plant or genetic stock for it's use at any time in future.

It is important to conserve the endangered plants or else some of the valuable genetic traits present in the existing and primitive plants will be lost. A global organization- International Board of Plant Genetic Resources (IBPGR) has been established for germplasm conservation and provides necessary support for collection, conservation and utilization of plant geneic resources through out the world.

The germplasm is preserved by the following two ways:

1. *In-situ conservation*: The germplasm is conserved in natural environment by establishing biosphere reserves such as national parks, sanctuaries. This is used in the preservation of land plants in a near natural habitat along with several wild types.
2. *Ex-situ conservation*: This method is used for the preservation of germplasm obtained from cultivated and wild plant materials. The genetic material in the form of seeds or in vitro cultures are preserved and stored as gene banks for long term use.

In vivo gene banks have been made to preserve the genetic resources by conventional methods *e.g.* seeds, vegetative propagules, etc. In vitro gene banks have been made to preserve the genetic resources by non - conventional methods such as cell and tissue culture methods. This will ensure the availability of valuable germplasm to breeder to develop new and improved varieties.

The methods involved in the in vitro conservation of germplasm are:

- *Cryopreservation*: In cryopreservation, the cells are preserved in the frozen state. The germplasm is stored at a very low temperature using solid carbon dioxide (at -79°C), using low temperature deep freezers (at -80°C), using vapour nitrogen (at -150°C) and liquid nitrogen (at-196°C). The cells stay in completely inactive state and thus can be conserved for long periods. Any tissue from a plant can be used for cryopreservation *e.g.* meristems, embryos, endosperms, ovules, seeds, cultured plant cells, protoplasts, calluses. Certain compounds like- DMSO (dimethyl sulfoxide), glycerol, ethylene, propylene, sucrose, mannose, glucose, praline, acetamide etc are added during the cryopreservation. These are called cryoprotectants and prevent the damage caused to cells (by freezing or thawing) by reducing the freezing point and super cooling point of water.
- *Cold Storage*: Cold storage is a slow growth germplasm conservation method and conserves the germplasm at a low and non-freezing temperature (1-9°C). The growth of the plant material is slowed down in cold storage in contrast to complete stoppage in cryopreservation and thus prevents cryogenic injuries. Long term cold storage is simple, cost effective and yields germplasm with good survival rate. Virus free strawberry plants could be preserved at 10°C for about 6 years. Several grape plants have been stored for over 15 years by using a cold storage at temperature around 9°C and transferring them in the fresh medium every year.
- *Low pressure and low oxygen storage*: In low- pressure storage, the atmospheric pressure surrounding the plant material is reduced and in the low oxygen storage, the oxygen concentration is reduced. The lowered partial pressure reduces the in vitro growth of plants. In the

low-oxygen storage, the oxygen concentration is reduced and the partial pressure of oxygen below 50 mmHg reduces plant tissue growth. Due to the reduced availability of O_2, and reduced production of CO_2, the photosynthetic activity is reduced which inhibits the plant tissue growth and dimension. This method has also helped in increasing the shelf life of many fruits, vegetables and flowers.

GENE TRANSFER METHODS IN PLANTS

To achieve genetic transformation in plants, we need the construction of a vector (genetic vehicle) which transports the genes of interest, flanked by the necessary controlling sequences *i.e.* promoter and terminator, and deliver the genes into the host plant. The two kinds of gene transfer methods in plants are:

VECTOR-MEDIATED OR INDIRECT GENE TRANSFER

Among the various vectors used in plant transformation, the Ti plasmid of Agrobacterium tumefaciens has been widely used. This bacteria is known as "natural genetic engineer" of plants because these bacteria have natural ability to transfer T-DNA of their plasmids into plant genome upon infection of cells at the wound site and cause an unorganized growth of a cell mass known as crown gall. Ti plasmids are used as gene vectors for delivering useful foreign genes into target plant cells and tissues. The foreign gene is cloned in the T-DNA region of Ti-plasmid in place of unwanted sequences.

To transform plants, leaf discs (in case of dicots) or embryogenic callus (in case of monocots) are collected and infected with Agrobacterium carrying recombinant disarmed Ti-plasmid vector. The infected tissue is then cultured (co-cultivation) on shoot regeneration medium for 2-3 days during which time the transfer of T-DNA along with foreign genes takes place. After this, the transformed tissues (leaf discs/calli) are transferred onto selection cum plant regeneration medium supplemented with usually lethal concentration of an antibiotic to selectively eliminate non-transformed tissues. After 3-5 weeks, the regenerated shoots are transferred to root-inducing medium, and after another 3-4 weeks, complete plants are transferred to soil following the hardening (acclimatization) of regenerated plants. The molecular techniques like PCR and southern hybridization are used to detect the presence of foreign genes in the transgenic plants.

VECTORLESS OR DIRECT GENE TRANSFER

In the direct gene transfer methods, the foreign gene of interest is delivered into the host plant cell without the help of a vector. The methods used for direct gene transfer in plants are: Chemical mediated gene transfer *e.g.* chemicals like polyethylene glycol (PEG) and dextran sulphate induce DNA uptake into plant protoplasts. Calcium phosphate is also used to transfer DNA into cultured cells. Microinjection where the DNA is directly injected into plant protoplasts or cells (specifically into the nucleus or cytoplasm) using fine tipped (0.5 - 1.0 micrometer diameter) glass needle or micropipette. This method of gene transfer is used to introduce DNA into large cells such as oocytes, eggs, and the cells of early embryo.

Electroporation involves a pulse of high voltage applied to protoplasts/cells/ tissues to make transient (temporary) pores in the plasma membrane which facilitates the uptake of foreign DNA.

The cells are placed in a solution containing DNA and subjected to electrical shocks to cause holes in the membranes. The foreign DNA fragments enter through the holes into the cytoplasm and then to nucleus.

Particle gun/Particle bombardment - In this method, the foreign DNA containing the genes to be transferred is coated onto the surface of minute gold or tungsten particles (1-3 micrometers) and bombarded onto the target tissue or cells using a particle gun (also called as gene gun/shot gun/ microprojectile gun).The microprojectile bombardment method was initially named as biolistics by its inventor Sanford. Two types of plant tissue are commonly used for particle bombardment- Primary explants and the proliferating embryonic tissues.

Transformation - This method is used for introducing foreign DNA into bacterial cells *e.g.* E. Coli. The transformation frequency (the fraction of cell population that can be transferred) is very good in this method. *E.g.* the uptake of plasmid DNA by E. coli is carried out in ice cold $CaCl_2$ (0-50C) followed by heat shock treatment at 37-450C for about 90 sec. The transformation efficiency refers to the number of transformants per microgram of added DNA. The $CaCl_2$ breaks the cell wall at certain regions and binds the DNA to the cell surface.

Conjuction - It is a natural microbial recombination process and is used as a method for gene transfer. In conjuction, two live bacteria come together and the single stranded DNA is transferred via cytoplasmic bridges from the donor bacteria to the recipient bacteria.

Liposome mediated gene transfer or Lipofection - Liposomes are circular lipid molecules with an aqueous interior that can carry nucleic acids. Liposomes encapsulate the DNA fragments and then adher to the cell membranes and fuse with them to transfer DNA fragments. Thus, the DNA enters the cell and then to the nucleus. Lipofection is a very efficient technique used to transfer genes in bacterial, animal and plant cells.

SELECTION OF TRANSFORMED CELLS FROMUNTRANSFORMED CELLS

The selection of transformed plant cells from untransformed cells is an important step in the plant genetic engineering. For this, a marker gene (*e.g.*

for antibiotic resistance) is introduced into the plant along with the transgene followed by the selection of an appropriate selection medium (containing the antibiotic). The segregation and stability of the transgene integration and expression in the subsequent generations can be studied by genetic and molecular analyses.

PROTOPLASTS FOR PLANT IMPROVEMENT

Protoplasts are naked cells that lack cell walls. They are spherical with a plasmolysed cell content and are contained within a plasmalemma. In principle, each individual protoplast can reform a cell wall, and later initiate either a callus through sustained divisions, or an embryo, defined as a somatic embryo. In banana they are obtained from in vivo tissues or in vitro cultures.

This phenomenon is termed totipotency and denotes the recovery of a whole organism from a single cell, and is also applicable to protoplasts. In theory all cells are totipotent, but in practice it depends on the past cellular environment. Usually the morphogenetic competence is retained at the unicellular stage that corresponds to a protoplast. However since protoplasts are not exposed to the stabilizing and inductive influence from neighbouring cells, they may have lost their plant regeneration capacity. In fact, factors such as the genotype or species of the manipulated plant, and the ontogenetic state of the explant source, exert a powerful effect on the regeneration potential of protoplasts.

Consequently the development of appropriate in vitro conditions for protoplast regeneration is complicated. Notwithstanding this, successful efforts have been made in isolation, cultivation and regeneration of protoplasts, since Nickell and Torrey pointed out their merit for crop improvement.

Various tissues and an increasing number of plant species and genotypes have been successfully used in protoplast culture, but so far sufficient quantities of protoplasts for practical applications are not routinely met. While the value for agriculture of protoplasts still needs to be demonstrated, they are an invaluable tool for studies on permeability of ions and solutes, photosynthesis, phytohormones, phytochrome, and maintenance of totipotency. Moreover, protoplasts are useful for the uptake of foreign genetic material and to produce somatic hybrids through protoplast fusion. In addition, protoplasts are an excellent system for studies on cell genetics and even for plant virology.

These studies rely on isolated, clean and healthy protoplasts, which requires the appropriate choice of osmoticum, hydrolysing enzymatic solution, and pretreatment of the donor explant either in vivo or in vitro. This involves the right choice of explant age, cold treatment, phytohormone pretreatment, light intensity or photoperiod, and subculture rhythm, which affects both the internal metabolic status of the cells and cell-wall composition. Prior to cultivation isolated protoplasts need to be freed from enzymatic remains and debris, which are considered to be toxic. Culture in or on solid medium is considered to be more advantageous than in liquid medium because the development of a single protoplast into a colony can be followed up much better. For development, different factors such as medium composition are crucial, especially the nature and quantity of growth hormones.

Other important factors are the physical environment, plating density, and embedding conditions or the use of feeder layers, which all contribute in one way or another to cell-wall regeneration and cell division. Once cell divisions start, the level of auxin(s), as well as the colony density, usually need to be reduced to avoid overcrowding. This is done by subculturing, which gives the additional benefit that an adequate nutrient supply is maintained. Finally, for rooting, plants are usually transferred to cytokinin-free medium, possibly containing some auxin, and exposed to high light intensities, usually in illuminated plant growth chambers.

During recent decades, plant biology workers have recognised the potential of protoplasts in many experimental systems, since an efficient enzymatic method for protoplast isolation was first established by Cocking.

SIGNIFICANCE AND USE OF PROTOPLASTSIN BANANA

Banana is now easily amenable to in vitro culture, and plants are regenerated from various explants through organogenesis, embryogenesis, anther culture and even from cultured protoplasts. This has created opportunities for other biotechnological applications. Cell suspensions, for example, opened the way for gene transfer and improved the production, the quality, and therefore the manipulation of cell suspension-derived protoplasts. Through genetic engineering it has become possible to confer new traits on banana plants, using either particle bombardment or Agrobacterium-mediated transfer.

Protoplasts facilitate the direct transformation of plant cells by DNA microinjection and electroporation. However, many characters of agricultural interest are multigenic or ill-defined, and current transformation methodologies allow the integration of only a few foreign genes. Protoplast fusion, however, allows the transfer of several useful characters, even if detailed genetic or molecular knowledge of genes encoding for these desired characters is lacking. Protoplast fusion is therefore a complementary tool to increase nuclear and cytoplasmic variability and to confer desirable agronomic traits.

Important banana cultivars are susceptible to many pests and diseases, particularly Mycosphaerella fijiensis and Fusarium oxysporum f.sp.cubense, nematodes and insects. Very interesting traits of resistance have been identified in this genus for most serious diseases. However cross-breeding is very difficult in the genus since most edible cultivars are sterile even after profuse pollination by wild pollen-fertile genotypes. Protoplast fusion is therefore an option, as it can overcome such sexual barriers which occur frequently in banana, and which cannot be bypassed even through embryo rescue. Therefore somatic hybridisation between wild and cultivated banana is expected to produce hybrids combining agronomic traits with genetic resistance to pests and pathogens.

Protoplast fusion has the added benefit that it creates the possibility of generating asymmetric fusions, whereby selected cytoplasmic organelles, chromosomes or chromosome fragments from an irradiated protoplast donor could be combined with the genome of an acceptor protoplast. This strategy is of especial interest in banana, where the subspecies balbisiana is considered to be a source of multiple resistance and therefore could be partially utilised to improve banana cultivars, of which the most important component is the subspeciesacuminata. Protoplast fusion also makes the creation of synthetic triploid banana genotypes possible by, for example, combining haploid protoplasts derived from anther culture with protoplasts from a diploid improved cultivar.

Chimerism and somaclonal variation are factors that seriously limit rapid clonal propagation of banana. By their nature, protoplasts avoid chimerism because they originate from a single cell, and protoplasts may be use to dissociate chimeric plants. On the other hand the cellular heterogeneity of protoplast populations can be useful for isolating somaclonal variants with improved characters such as increased yield or pathogenic resistance, so that somaclonal variation may also be exploited to provide new sources of genetic variability.

Banana protoplast isolation and culturing is nowadays routine, although protoplast research on banana started 15 years later than on model plants. Now it is regrettably underused because of the strong emphasis on molecular and genomic studies in banana. Nevertheless, protoplasts have much to offer for non-conventional banana breeding, since they overcome sexual incompatibilities at interspecific and even at the intergeneric level, and allow the incorporation of multigenic traits such as yield, resistance to stress, pests and diseases. Finally protoplast techniques may improve banana when other classical methods have failed.

WORK ON ISOLATED CELLS AND PROTOPLASTS

Early Experiments

The first attempts at banana protoplast culture were made in 1984 by Bakry. Various explants (leaf and sheath from in vitro banana plants, callus from floral explants, and bract and tepals from in vivo banana plants) of diploid and triploid banana plants belonging to the AA, BB, AAA and AAB groups were tested.

Enzyme solutions were composed of a modified CPW (Cell and protoplast washing solution) salt solution containing 0.7 M mannitol, and a mixture of various enzymes at 0.1 to 5% (w/v) such as:

- *Pectinases*: Macerozyme;
- *Hemicellulases*: Hemicellulase
- *Cellulases*: cellulase R-10 'Onozuka', Cellulysin and Driselase.

Table. Cell and Protoplast Washing (CPW) Solution

Component	Concentration (mg/l)
KH_2PO_4	27.2
KNO_3	100
$CaCl_2.2H_2O$	150
$Mg\ SO_4.7H_2O$	250
KI	0.16
$Cu\ SO_4.5H_2O$	0.025

The enzyme solution was adjusted to pH 5.6 with KOH and filter-sterilised (0.22 mm). Small explants of about 1-5 mm^3 were incubated for 9 h with shaking (80 r.p.m.) at 27°C in the dark. The mixture was sieved through an 83 mm metallic sieve, followed by dilution with KMC salt solution. After centrifugation at 100 g for 5 min, the pellet was diluted in a KMC solution. Whatever the plant material, enzymatic solution and plant preconditioning (light/dark), only isolated plasmolysed cells were obtained. Positive results were obtained when nodular calli were used as starting material. Floral explants cultivated on MS medium complemented with 500 mg/l casein hydrolysate, 2 mg/l IAA, and 2 mg/l BAP lead to calli covered with nodular bodies. These calli were embryogenic. With the enzyme mixture 2.5% cellulase R10 'Onozuka', 0.2% hemicellulase, 0.3% pectolyase Y23 and 0.6% macerozyme, Bakry produced the first living banana protoplasts. After calcofluor staining, protoplasts were found to be cellulose-free, perfectly round in shape, variable in size and having a dense cytoplasmic content. Although the yield was quite low the protocol was reproducible.

This pioneering work was confirmed by Cronauer and Krikorian in 1986 (using proliferating shoot tips of the cultivar 'Lacatan' [AAA group]) and by Da Silva Conceicao in 1989. Later on regular progress was made by several scientists. Initially banana calli from protoplasts were developed in 1992, followed by plant regeneration directly from protoplasts, protoplast transformation and somatic hybridisation.

Explants used as a Source of Protoplasts

In banana, in vivo explants are not a good source of protoplasts. Although Matsumoto *et al.* obtained protoplasts from in vivo bracts, in most cases initiation started from in vitro cultures. Leaf explants, slices of shoot tissue, roots or callus were the first choice as starting material for protoplast isolation because of their convenience. In fact, protoplasts can be obtained in banana from almost any tissue, including young leaves, sheaths, bracts, roots, and callus, but yields depend on the explant source. Yields are quite low with most tissues and range from 0.1 to 2.8 × 106 protoplasts per gram, but the protoplasts are unable to divide. However, cell suspension-derived protoplasts gave high yields which ranged from 4 to 5 × 105 per gram with 'Long Tavoy'; from 1.1 to 3 × 105per gram with malaccensis; 2 to 20 × 106 per gram with 'Maça' to 6.6 × 107 per millilitre of packed cell volume (PCV) with 'Bluggoe'. Essential was the embryogenicity of the suspension. The first sustained cell divisions in protoplast culture leading to calluses were observed by Megia *et al.* in 1992 with M. acuminata cell suspension-derived protoplasts. Megia *et al.* and Panis *et al.* improved the protocol. This was confirmed by Matsumoto *et al.* and Assani *et al.* who produced plants from such 'embryogenic protoplasts'. Consequently embryogenic cell suspensions are the starting material of choice.

Banana Genotypes Tested

Bananas are either triploid, diploid or tetraploid in order of importance for food production, but protoplasts have been obtained so far from triploid, diploid and haploid genotypes. Below are listed the cultivars and wild types tested; the letter A denotes the contribution of the subspecies acuminata and B the contribution of the subspecies balbisiana.

Triploid Genotypes

- Musa cavendishii L.cv North banana (AAA);
- 'Lacatan' (AAA)
- 'Nanicao' (AAA)
- 'Bluggoe' (ABB)
- 'Maça' (AAB); 'Grande Naine' (AAA); 'Gros Michel' (AAA); 'Currare Enano' (AAB); 'Dominico' (AAB).

Diploid Genotypes

- 'Pisang Lilin' (AA)
- Musa acuminata ssp. burmanica type 'Long Tavoy' (AA)
- Musa acuminata ssp. malaccensis (AA)
- SF 265 (AA); IRFA 903 (AA); Col 49 (AA); 'Colatino Ouro' (AA); 'Pisang Klutuk Wulung' (BB); 'Pisang Klutuk' (BB); 'Pisang Batu' (BB) and 'Tani' (BB).

Protoplast Extraction and Purification

Enzymes

Various enzyme mixtures are used depending on the author and banana explant (CS = Cell Suspension; P = Plant explant; C = callus).

They are commonly composed of:

- Cellulases (Cellulase Onozuka R10, Cellulase Onozuka RS, Cellulysin, Macerase, Driselase)
- Pectinases (Pectinase; Macerozyme R10; Pectolyase Y 23)

- Hemicellulase (Rhozyme HP -150; Glusulase)

Protoplast Isolation Media

It is standard practice to incubate the plant material in the enzyme solution for a digestion period of 1-24 h either with shaking (40 r.p.m.) or no shaking, generally at 27°C in the dark. The best source of protoplasts is from embryogenic cell suspension cultivated in the light (30 or 65 mE m^{-2} s^{-1}) or darkness and regularly subcultured. Initiation should start 3-5 days, 7-10 days or 2-15 days after the last subculture, followed by sieving through a 200-400 mm mesh just before enzyme incubation.

Protoplast Purification

Protoplasts are purified from debris by sieving (100 mm and subsequently or directly through 56/32/25 mm sieves), which is preferred to flotation. With flotation purification, a 21% sucrose solution and centrifugation at 120 g is used.

Protoplast Washing

After sieving, protoplasts are generally repeatedly washed to remove enzymes. Protoplast pellets obtained by centrifugation (50/66/90 g) are washed with protoplast isolation medium without enzymes; cell suspension culture medium without growth regulator plus 10% mannitol; or 3% $CaCl_2$ plus 0.5% KCl solution; or 3% $CaCl_2$ 0.5% KCl solution and 10% mannitol; or 0.6 M D-mannitol, 0.1 mM $CaCl_2$, 0.5% PVP-40 and 3.5 mM MES.

Protoplast Quality Control

Complete cell wall degradation is confirmed both by the spherical shape of the released protoplasts and by the absence of fluorescence after staining with calcofluor white. Evans blue or FDA staining are commonly used for viability assessment.

Protoplast Culture Media

For banana protoplast culture, derived MS medium (supplemented with 2,4-D, or 2,4-D plus zeatin) or N-medium composed with N6 salt, vitamins, organic acid and sugar alcohol are commonly used. Due to difficulties in obtaining favourable protoplast development, many culture systems have been tested.

Liquid and solid media (derived from MS) were initially tried for protoplast cultivation, but it was found that liquid cultures are not suitable for Musa protoplasts. In contrast, solid media supported sustained divisions.

To overcome the low growth response, other media were tested, such as semi-solid media, protoplasts embedded in solid or semi-solid media, nurse cultures, and nurse cultures with a feeder layer. Cocultivation in banana protoplast cultures was considered as it was reported that coculturing protoplasts of recalcitrant species, especially monocotyledons, with a reliable feeder culture, induced cell division. The banana protoplast literature describes various combinations of media and culture systems but, whatever the protocol, a high plating density (10^5-10^6/ml) was crucial, as well as the use of feeder layer, or nurse culture combined with a feeder layer. Protoplasts are cultured at 27 ± 1°C in the dark.

Culture in Liquid Media

For culturing protoplasts in liquid media, protoplasts with a density of 105-106 per millilitre are suspended in liquid media, with or without shaking. The following options are open:

- N6 salt and KM vitamins, organic acids and sugar alcohols according to Kao and Michayluk, 6.3% (w/v) glucose and 4% (w/v) sucrose as osmoticum, 260 mg/l KH_2PO_4 and 100 mg/l 2-(N-morpholino) ethanesulphonic acid MES;
- N_6 salts, KM vitamins, organic acids and sugar alcohols, vitamins of Morel, 0.4 M glucose, 117 mM sucrose, 0.5 mM MES, 1.9 mM KH_2PO_4;
- N_6 salts, KM vitamins, organic acids and sugar alcohols, vitamins of Morel, 0.4 M glucose, 117 mM sucrose, 1.9 mM KH_2PO_4, 2.28 μM zeatin, 0.90 μM 2,4-D and 5.4 μM NAA;
- MS modified medium, 5 mM 2,4-D and 10% mannitol;
- 1/2 MS, 10% mannitol;
- MS medium, 5 mM 2,4-D, 1 mM zeatin, 0.55 M mannitol;
- MS minor nutrients, 1/2 major nutrients, MS vitamins, 10 mg/l ascorbic acid, 20 g/l sucrose, 5 μM 2,4-D; 0.55 M D-mannitol;
- MS preconditioned medium, 5 mM 2,4-D, 10% filter-sterilised medium from a one-week-old cell culture.

Culture on Solid Medium

For culture on solid medium protoplasts are directly plated at high density on a gelatinous medium or on a membrane which covers the gelatinous medium. *Such media are composed of*:

- MS minor nutrients, 1/2 major nutrients, MS vitamins, 10 mg/l ascorbic acid, 20 g/l sucrose, 5 μM 2,4-D, 0.275 M D-mannitol, 0.2% gelling gum; or
- MS 1/2 salt solution, 5% mannitol, 5 μM 2,4-D, 0.8% agarose or 0.2% Gelrite.

Culture in Semi-solid Medium

For culture on semi-solid medium protoplasts are cultivated on:

- 0.6% agarose, N_6 salt, vitamins of Morel, vitamins of Kao and Michayluk;

- 0.6% agarose, 0.35 M glucose, 40 g/l saccharose, 260 mg/l KH_2PO_4, 3 mM MES, and Pichloram, BAP, NAA, L-asparagine and L-proline at various concentrations;
- Protoplasts at twice final desired concentration in culture medium, are mixed with the same medium containing 0.3/0.8% agarose, 5-10% mannitol, 0-10 μM 2,4-D.

Culture Embedded in Solid Medium

For embedment in solid medium protoplasts at twice the desired concentration (2×10^6) are placed in a twofold concentrated culture medium: MS minor nutrients (× 2), MS major nutrients, MS vitamins (× 2), 20 mg/l ascorbic acid, 40 g/l sucrose, 10 μM 2,4-D, but with 0.55 M D-mannitol. This is followed by mixing with an equal volume of solution containing a gelling gum at 0.6%, 0.55 M D-mannitol, previously melted and maintained at 55°C.

Culture in Alginate

Protoplasts are cultured in a solid medium, or in calcium alginate beads according to Schilde-Rentschler *et al*. For this purpose, the alginate solution (2.8% alginic acid, 0.4 M mannitol) is first autoclaved. Protoplasts are added to the alginate solution to obtain a concentration of 1.0×10^6 protoplasts per millilitre.

Later, drops of the protoplast-alginate mixture are polymerised in 0.4 M mannitol/50 mM CaCl2 solution, and transferred into protoplast liquid culture medium in small Petri dishes. Pieces of solid medium containing protoplasts are transferred into liquid medium:

- MS medium, 10 mg/l ascorbic acid, 20 g/l sucrose, 5 μM 2,4-D, 0.275 M D-mannitol, with or without 0.3% activated charcoal;
- 'A medium' consisting of N6 salts, KM vitamins, organic acids and sugar alcohols, vitamins of Morel, 117 mM sucrose, 1.9 mM KH_2PO_4;
- 'B medium' consisting of the 'A medium' containing 2.28 μM zeatin, 0.90 μM 2,4-D and 5.4 μM NAA.

Nurse Cultures

Living cells as nurse in liquid medium: Protoplasts can be maintained according to one of the following procedures:

- Nurse cell suspensions are maintained in N6 salt with MS vitamins, 10 μM L-proline, 300 mg/l casein hydrolysate, 5 μM 2,4-D, and 3% (w/v) sucrose at pH 5.8;
- 1 ml PCV of nurse cells is freshly subcultured and mixed with 50 ml of MS minor nutrients, 1/2 major nutrients, MS vitamins, 10 mg/l ascorbic acid, 20 g/l sucrose, 5 μM 2,4-D and 0.275 M D-mannitol. Embedded protoplasts prepared as above are then immersed in the nurse solution. Cultures are placed on a rotary shaker at 120 r.p.m.

A MillicellTM-HA μm pore size) may be used as a physical barrier, to separate protoplasts from nurse cells during culture.

Feeder-layer Culture of Living Cells

For banana protoplast culture, Oryza, Lolium multiflorum, Triticum monococcum (Poaceae) and banana cell suspensions are useful as nurse cells. Feeder cultures are prepared one day before protoplast isolation by mixing 3 ml of feeder cell suspension with 100 ml of a medium containing 0.8% (w/v) melted Sea plaque agarose, MS basal salt, vitamins of Morel, 2 mg/l 2,4-D, 20 g/l sucrose, 72 g/l maltose, 250 mg/l glucose and 20 ml/l fresh coconut water. Feeder cells are then separated from overlaying protoplasts by a 10 mm nylon filter or 0.22 mm Millipore membrane.

Others options are:

- 2 ml of a nurse culture are poured on 20ml 1/2 MS, 5% mannitol, 5 μM 2,4-D, 0.2% Gelrite. The nurse culture is made from 1 ml of one-week-old suspension concentrated to 25% PCV mixed with 1 ml 1/2 MS supplemented with 10% mannitol, 200 mg/l myoinositol, 5 μM 2,4-D and 1.6% agarose.
- Nurse cells are mixed in melted medium MS (but with major nutrients at half strength), 10 mg/l ascorbic acid, 5 μM 2,4-D, sucrose 20 g/l, 0.55 M D-mannitol, 0.8% agarose (plaque).
- 1 or 2 days before protoplast culture, prepare the feeder layer.

The cell suspensions are sieved through a 250 μm nylon filter to select small cell aggregates.

The PCM (Protoplast Culture Medium) liquid medium, which consists of MS salts, 9 μM 2,4-D, vitamins of Morel, 2.8 mM glucose, 278 mM maltose, 116 mM sucrose, 2.5 mM myoinositol (pH 5.7) is sterilised through 0.22 mm Millipore Millex GS filters. Nurse cell suspensions are added to 100 ml double-concentration PCM culture medium to obtain a final cell concentration of 3-10% (v/v). 1.2 g agarose Sea plaque (Sigma) is dissolved in 100 ml water and then autoclaved (pH 5.7); when the agarose solution has cooled to 30-35°C, it is gently mixed with 100 ml PCM medium containing nurse cells, and 10-12 ml of the mixture is poured into small Petri dishes (5.5 cm diameter). After solidification, the medium is covered with a sterilised nitrocellulose filter (AA type, Millipore Corporation). 0.5 ml of the protoplast suspension in the A or B medium is transferred onto the nitrocellulose filter. Cell-wall regeneration is observed with calcofluor white. All cultures (liquid culture, alginate culture and feeder-layer culture) are maintained at 27°C in the dark.

- *Culture on Preconditioned Medium*: This cultivation technique resembles the feeder-layer technique, except that the preconditioned medium is the cell culture filtered through a Whatman filter, followed by mixing the filter-sterilised medium with the solidifying medium.

Plant Regeneration

Sustained cell divisions are obtained with the feeder-layer technique, leading to plant development, while other techniques give only protoplast budding or few divisions. A high protoplast density (10^5-10^6 per millilitre) also appears to be crucial for plant regeneration. After 3-8 days of culture, first and second divisions are observed. After 4-6 weeks of cultivation, cell clumps consist of cells with embryogenic characters (*i.e.* small cell size, large nucleus with very stainable nucleolus, dense cytoplasm, small vacuoles).

Since the clumps all have a similar size, it seems that they all arise at the same time from the protoplasts. For plant regeneration, cell clumps are transferred to a semi-solid medium without nurse cells, devoid of mannitol and growth regulators, or are transferred onto a medium containing 2.2 mM BAP and 2 mM pichloram for one week before transfer to MS medium containing 10 mM BAP to improve the rate of conversion of globules into plants.

The membrane on which one-month-old clusters are growing is transferred onto solid medium with minor MS nutrients, 1/2 major MS nutrient strength, MS vitamins, 10 mg/l ascorbic acid, 20 g/l sucrose, 5 μM 2,4-D, 0.275 M D-mannitol, 0-2% gelling gum to produce embroids. Their maturation is achieved by transfer on hormone-free medium without mannitol, and finally on MS solid medium without growth regulators. Plantlet regeneration increased in the presence of BAP and IAA at low concentrations (2 mM) before transplanting to the soil in the greenhouse.

BIOTECHNOLOGY IN PLANT CELL

Clonal propagation refers to the process of asexual reproduction by multiplication of genetically identical copies of individual plants. The vegetative propagation of plants is labour-intensive, low in productivity and seasonal. The tissue culture methods of plant propagation, known as 'micropropagation' utilizes the culture of apical shoots, axillary buds and meristems on suitable nutrient medium.The regeneration of plantlets in cultured tissue was described by Murashige in 1974. Fossard gave a detailed account of stages of micropropagation.

The micropropagation is rapid and has been adopted for commercialization of important plants such as banana, apple, pears, strawberry, cardamom, many ornamentals (*e.g.* Orchids) and other plants.

The micropropagation techniques are preferred over the conventional asexual propagation methods because of the following reasons: (a) In the micropropagation method, only a small amount of tissue is required to regenerate millions of clonal plants in a year., (b) micropropagation is also used as a method to develop resistance in many species., (c) in vitro stock can be quickly proliferated as it is season independent,. (d) long term storage of valuable germplasm possible.

The steps in micropropagation method are:

- *Initiation of culture*: From an explant like shoot tip on a suitable nutrient medium,
- Multiple shoots formation from the cultured explant,
- Rrooting of in vitro developed shoots and,
- Transplantation - transplantation to the field following acclimatization.

The factors that affect micropropagation are:

- Genotype and the physiological status of the plant *e.g.* plants with vigorous germination are more suitable for micropropagation.,
- The culture medium and the culture environment like light, temperature etc. For example an illumination of 16 hours a day and 8 hours night is satisfactory for shoot proliferation and a temperature of 25°C is optimal for the growth.

The benefits of micropropagation this method are:

- Rapid multiplication of superior clones can be carried out through out the year, irrespective of seasonal variations.
- Multiplication of disease free plants *e.g.* virus free plants of sweet potato (Ipomea batatus), cassava (Manihot esculenta)
- Multiplication of sexually derived sterile hybrids
- It is a cost effective process as it requires minimum growing space.

SOMACLONAL VARIATION

The genetic variations found in the in vitro cultured cells are collectively referred to as somaclonal variation and the plants derived from such cells are called as 'somaclones'. It has been observed that the long-term callus and cell suspension culture and plants regenerated from such cultures are often associated with chromosomal variations. It is this property of cultured cells that finds potential application in the crop improvement and in the production of mutants and variants (*e.g.* disease resistance in potato).

Larkin and Scowcroft working at the division of Plant Industry, C.S.I.R.O., Australia gave the term 'somaclones' for plant variants obtained from tissue cultures of somatic tissues. Similarly, if the tissue from which the variants have been obtained is having gametophytic origin such as pollen or egg cell, it is known as 'gametoclonal' variation.

They explained that it may be due to:

- Reflection of heterogeneity between the cells and explant tissue,
- A simple representation of spontaneous mutation rate, and
- Activation by culture environment of transposition of genetic materials.

Shepard *et al.* also contributed by screening about 100 somaclones produced from leaf protoplasts of Russet Burbank. They found that there was a significant amount of stable variation in compactness of growth habit, maturity, date, tuber uniformity, tuber skin colour and photoperiodic requirements.Somaclonal Variations has been used in plant breeding programmes where the genetic variations with desired or improved characters are introduced into the plants and new varieties are created that can exhibit disease resistance, improved quality and yield in plants like cereals, legumes, oil seeds tuber crops etc. Somaclonal variation is applicable for seed

APPLICATIONS OF SOMACLONAL VARIATIONS

- Methodology of introducing somaclonal variations is simpler and easier as compared to recombinant DNA technology.
- Development and production of plants with disease resistance *e.g.* rice, wheat, apple, tomato etc.
- Develop biochemical mutants with abiotic stress resistance *e.g.* aluminium tolerance in carrot, salt tolerance in tobacco and maize.
- Development of somaclonal variants with herbicide resistance *e.g.* tobacco resistant to sulfonylurea
- Development of seeds with improved quality *e.g.* a new variety of Lathyrus sativa seeds (Lathyrus Bio L 212) with low content of neurotoxin.
- Bio-13 – A somaclonal variant of Citronella java (with 37% more oil and 39% more citronellon), a medicinal plant has been released as Bio-13 for commercial cultivation by Central Institute for Medicinal and Aromatic Plants.
- Supertomatoes- Heinz Co. and DNA plant Technology Laboratories (USA) developed Supertomatoes with high solid component by screening somaclones which helped in reducing the shipping and processing costs.

Production of Virus Free Plants

The viral diseases in plants transfer easily and lower the quality and yield of the plants. It is very difficult to treat and cure the virus infected plants therefore te plant breeders are always interested in developing and growing virus free plants.In some crops like ornamental plants, it has become possible to produce virus free plants through tissue culture at the commercial level.

This is done by regenerating plants from cultured tissues derived from:

- Virus free plants,
- Meristems which are generally free of infection - In the elimination of the virus, the size of the meristem used in cultures play a very critical role because most of the viruses exist by establishing a gradient in plant tissues. The regeneration of virus-free plants through cultures is inversely proportional to the size of the meristem used.,
- Meristems treated with heat shock (34-36°C) to inactivate the virus,

- Callus, which is usually virus free like meristems.
- Chemical treatment of the media- attempts have been made to eradicate the viruses from infected plants by treating the culture medium with chemicals *e.g.* addition of cytokinins suppressed the multiplication of certain viruses.

Among the culture techniques, meristem-tip culture is the most reliable method for virus and other pathogen elimination.

Viruses have been eliminated from a number of economically important plant species which has resulted in a significant increase in the yield and production *e.g.* potato virus X from potato, mosaic virus from cassava etc.

These virus free plants are not disease resistant so there is a need to maintain stock plants to multiply virus free plants whenever required.

Production of Synthetic Seeds

In synthetic seeds, the somatic embryos are encapsulated in a suitable matrix (*e.g.* sodium alginate), along with substances like mycorrhizae, insecticides, fungicides and herbicides. These artificial seeds can be utilized for the rapid and mass propagation of desired plant species as well as hybrid varieties.

The major benefits of synthetic seeds are:

- They can be stored up to a year with out loss of viability
- Easy to handle and useful as units of delivery
- Can be directly sown in the soil like natural seeds and do not need acclimatization in green house.

Mutant Selection

An important use of cell cultures is in mutant selection in relation to crop improvement. The frequency of mutations can be increased several fold through mutagenic treatments and millions of cells can be screened. A large number of reports are available where mutants have been selected at cellular level. The cells are often selected directly by adding the toxic substance against which resistance is sought in the mutant cells. Using this method, cell lines resistant to amino acid analogues, antibiotics, herbicides, fungal toxins etc have actually been isolated.

Production of Secondary Metabolites

The most important chemicals produced using cell culture are secondary metabolites, which are defined as' those cell constituents which are not essential for survival'. These secondary metabolites include alkaloids, glycosides (steroids and phenolics), terpenoids, latex, tannins etc. It has been observed that as the cells undergo morphological differentiation and maturation during plant growth, some of the cells specialize to produce secondary metabolites. The in vitro production of secondary metabolites is much higher from differentiated tissues when compared to non-differentiated tissues. The cell cultures contribute in several ways to the production of natural products.

These are:

- A new route of synthesis to establish products *e.g.* codeine, quinine, pyrethroids,
- A route of synthesis to a novel product from plants difficult to grow or establish *e.g.* thebain from Papaver bracteatum,
- A source of novel chemicals in their own right *e.g.* rutacultin from culture of Ruta,
- As biotransformation systems either on their own or as part of a larger chemical process *e.g.* digoxin synthesis.

The Advantages of In Vitro Production of Secondary Metabolites

- The cell cultures and cell growth are easily controlled in order to facilitate improved product formation.
- The recovery of the product is easy.
- As the cell culture systems are independent of environmental factors, seasonal variations, pest and microbial diseases, geographical location constraints, it is easy to increase the production of the required metabolite.
- Mutant cell lines can be developed for the production of novel and commercially useful compounds.
- Compounds are produced under controlled conditions as per the market demands.
- The production time is less and cost effective due to minimal labour involved.

4

The Role of Plant Biotechnology in the World's Food Systems

At the molecular level, different organisms are quite similar, writes Cornell University Professor A.M. Shelton. It is this similarity that allows the transfer of genes of interest to be moved successfully between organisms, therefore, genetic engineering is a much more powerful tool than traditional breeding in improving crop yields and promoting environmentally friendly production methods. For the past 10,000 years, humans have used the plants nature provided and modified them through selective breeding to have desirable characteristics such as improved taste, enhanced yield and pest resistance.

The result is that the plants we consume today would be largely unrecognizable to our ancient ancestors. Scientists consider the techniques of biotechnology to be an aid in the selective breeding of plants and to have far more potential for providing benefits such as enhanced nutritional properties, more environmentally friendly production methods and improved yields. Already, the techniques of biotechnology have produced tremendous benefits in medicine. Virtually all the insulin used to treat diabetes today is produced through biotechnology and genetic engineering, and many of the medicines used to fight cancers and heart problems are produced through these same methods.

BIOTECHNOLOGY IN THE GLOBAL COMMUNICATION ECOLOGY

Public debates about the safety of new products introduced in the market go back centuries and were often based less on science than on the politics of the time. Similarly, today, much of the debate about agricultural biotechnology is steered by myths and misinformation and not by science, writes Calestous Juma, professor and director of the Science, Technology and Globalization Project at the Kennedy School of Government, Harvard University. The scientific community, with stronger support from governments, must do more to address science and technology issues with the public, he adds. Debates over biotechnology are part of a long history of social discourse over new products. Claims about the promise of new technology are at times greeted with skepticism, vilification or outright opposition—often dominated by slander, innuendo and misinformation.

Even some of the most ubiquitous products endured centuries of persecution. For example, in the 1500s Catholic bishops tried to have coffee banned from the Christian world for competing with wine and representing new cultural as well as religious values. Similarily, records show that in Mecca, in 1511 a viceroy and inspector of markets, Khair Beg, outlawed coffeehouses and the consumption of coffee.

He relied on Persian expatriate doctors and local jurists who argued that coffee had the same impact on human health as wine. But the real reasons lay in part in the role of coffeehouses in eroding his authority and offering alternative sources of information on social affairs in his realm. In public smear campaigns similar to those currently directed at biotech products, coffee was rumored to cause impotence and other ills and was either outlawed or its use restricted by leaders in Mecca, Cairo, Istanbul, England, Germany and Sweden. In a spirited 1674 effort to defend the consumption of wine, French doctors argued that when one drinks coffee: "The body becomes a mere shadow of its former self; it goes into a decline, and dwindles away. The heart and guts are so weakened that the drinker suffers delusions, and the body receives such a shock that it is as though it were bewitched."

BUTTERFLY STORIES AND OTHER MISINFORMATION TACTICS

Today similarly charged stories are told about genetically modified (GM) foods. In addition to claims about the negative impact of GM foods on the environment and human health, there are wild claims that associate GM foods with maladies such as brain cancer and impotence as well as behavioural changes. Some of these rumors are spread at the highest levels of government in developing countries.

The tactics employed in the debates are equally sophisticated. Critics of the technology have used instruments of mass communication to provide the public with information that is carefully designed to highlight the dangers they attribute to biotechnology. Advocates of biotechnology have often been forced to respond to charges against the technology and have only on rare occasions taken the initiative to reach out to the public.

This is particularly important because the general public does not readily understand the technical details of biotechnology products and so new communication approaches are needed. While advocates of biotechnology have often tried to rely on the need for scientific accuracy, critics employ rhetorical methods that are designed to invoke public fear and cast doubt on the motives of the industry. The critics draw analogies between the "dangers" of biotechnology with the catastrophic consequences of nuclear power or chemical pollution. Indeed, they use terms like "genetic pollution" and "Frankenstein foods". Critics have also relied on the general distrust of large corporations among sections of the global community to make their case. In addition, they have made effective use of incidents, whose risks they have amplified.

A muchquoted study by Cornell University researchers indicated that pollen from GM maize producing a Bt toxin killed the larvae of Monarch butterflies. This study was used to dramatize the impact of biotechnology on the environment. Subsequent published peer explanations of the limitations of the study and refutations of the conclusions did not change the original impression created by the critics of biotechnology.

In this case the real environmental issue was not if GM maize killed monarch butterfly larvae or not. The critical question was what impact the maize had on the environment compared to maize grown with chemical pesticides. It is the issue of relative risks that is important; not simply a single event examined outside the wider ecological context. But apparently, this kind of analysis would not serve the cause of critics.

It is notable that the critics of biotechnology have defined the rules of the debate in two fundamental ways. First, they have managed to create the impression that the onus of demonstrating safety lies with advocates of biotechnology and not on its critics. In other words, biotechnology products are considered unsafe until proven otherwise. Second, they have been effective in framing the debate in environmental, human health and ethical terms, thereby masking the underlying international trade considerations. By doing so, they have managed to rally a much wider constituency of activists who are genuinely concerned about environmental protection, consumer safety and ethical social values.There is a general view that concerted efforts to promote public debate will improve communication and lead to the acceptance of biotechnology products. This may be the case in some situations. But generally, the concerns are largely material and cannot be resolved through public debate alone. This is mainly because the root causes of the debate lie in the socio-economic implications of the technology and not mere rhetorical considerations.

It is possible that public debates will only help to clarify or amplify points of divergence and do little to address fundamental economic and trade issues. What then can be done under the circumstances, especially in relation to developing countries that are currently the target of much of the attention of advocates and critics of biotechnology? Operating in the new global communication ecology will require greater diversity of biotechnology products, an increase in the number of institutional players, enhanced policy research on life sciences and society, and stronger policy leadership.

Products Speak Louder than Words

Much of the debate on the role of biotechnology in developing countries is based on hypothetical claims with no real products in the hands of producers or consumers. Under such circumstances, communication and dialogue are not enough until there is a practical reference point. In other words, rebutting the claim of critics is not as important as presenting the benefits of real products in the market place.

This can best be achieved through collaborative efforts among local scientists, entrepreneurs, policy makers and legitimate civil society organizations. There is ample evidence to suggest that concerns over the safety of new products tend to decline as local participation and ownership in new technologies increase. Similarly, local participation in new technologies increases the level of trust in new technologies, thereby reducing the demand for non-science-based safety regulations. For example, the word of a farmer from South Africa stating the positive impact of GM cotton on her welfare carries more weight than thousands of screaming press releases and empty headlines on both sides of the debate. This means that spreading the use of biotechnology not only promotes familiarity with the technology, but also generates the information needed to convince the public about the relevance and usefulness of the technology.

The broadening of the range of products is therefore a key aspect of the debate. This is particularly important in developing countries interested in using the technology to enhance local products and diversify their food base. Information on the development of drought-tolerant crops, for example, would be relevant to African countries while other regions might be interested in different products. This view also suggests that general debates about the role of biotechnology are of little utility unless framed in the context of local needs and applications. The absence of a real stake in the technology creates a vacuum that is often filled with misinformation on the risks and benefits of the technology. Countries such as Kenya and South Africa that have their own biotechnology research programmes have a more considered view of the technology.

Broadening the Constituency

Addressing the issue of biotechnology communication requires an improved understanding of the changing ecology of communication. The ecology includes a complex network of sources of information and opinion leaders as well as new communication tools that were hitherto not available to the public or advocacy groups.

In his days, Khair Beg was outraged to learn that coffeehouses had become an authoritative source of information on what was happening in his jurisdiction. Similarly, the Internet has become a more important communication tool than classical methods such as TV advertising. But unlike in the days of Khair Beg, the new communication ecology is global in character, making it possible to spread information widely and generate empathy among a diversity of activist organizations, including those that are unlikely to be affected by the technology.

These cyber-communities are built around a complex set of mailing lists that are not easily accessible. Correcting misinformation spread through such channels is difficult to do because of the complexity of the networks. While activists tend to use a diverse array of social movements to advance their cause, advocates have tended to focus on the use of centralized institutions whose impact is largely negligible in the modern communication ecology. But creating the necessary diversity requires a broadening of the base of social movements that champion the role of science and technology in human welfare. One of the most important aspects of the biotechnology debate has been the role of the popular media. In Europe, for example, the media have played an important role in amplifying claims by critics or creating doubt about positions advanced by advocates of the technology.

In contrast, support for the role of science does not usually have the polemical turn that newspaper editors relish. The traditional view that science is based on immutable facts which can be passed on from an authority to the general public is being challenged by approaches that demand greater participation in decision-making.

In other words, scientific information is being subjected to democratic practices. The debate over biotechnology has pushed the frontiers of public discourse of technical matters. On the one hand, society is being forced to address issues that are inherently technical, and on the other, the scientific community is under pressure to accept non-technical matters as valid inputs to decision-making.

Thinking Ahead

Policy-oriented research institutions and think tanks play an important role in the war of words. It is notable that critics of biotechnology have made a considerable effort to create alliances with research institutions, including university-based departments. Much of the material used to question the safety of biotechnology often has the legitimacy of a research institution.

But non-partisan policy research on the role of biotechnology in society is largely lacking, and so those seeking to provide an alternative view have limited opportunities to obtain credible information.

The lack of systematic research on the interactions between biology and society is a critical bottleneck in efforts to engage the public in dialogue on biotechnology. This is particularly critical given the fact that advances in biology pose new ecological and ethical issues that are not associated with the physical and chemical sciences. For example, concerns over the inability to recall products once released on markets are more pronounced when dealing with the release of biological inventions into the environment.

Leading the Way

Much of the public debate is intended to influence government policy on biotechnology. In this regard, the capacity of governments to assess the available information and use it for decision-making is an essential element of the debate. Political leadership on biotechnology and the existence of requisite institutions of science and technology advice are an essential aspect of the governance of new technologies. Debates over new technologies will be more pronounced in the future, and governments will increasingly come under pressure to address these issues.

But science and technology advice will not be sufficient unless governments view science and technology as integral to the development process. In this regard, enhancing the capacity of leadership to address science and technology issues will contribute to the effective management of public debates over new technologies in general and biotechnology in particular. On the whole, the nature of emerging technologies—particularly those based on the life sciences—and the changing ecology of communication are making it necessary to rethink strategies for advancing the role of biotechnology in society.

The scientific community will need not only to demonstrate a clear sense of leadership, but also to adapt its communication methods to suit the growing complexity and diverse needs of the global community. In the final analysis, it is the range of useful products available to humanity from biotechnology that will settle the debate, not the hollow pronouncements of advocates and critics.

THE GENE REVOLUTION: A CHANGING PARADIGM FOR AGRICULTURAL R&D

The Green Revolution brought high-yielding semi-dwarf wheat and rice varieties, developed with conventional breeding methods, to millions of small-scale farmers, initially in Asia and Latin America, but later in Africa as well. The gains achieved during the early decades of the Green Revolution were extended in the 1980s and 1990s to other crops and to less favoured regions. In comparison with the research that drove the Green Revolution, the majority of agricultural biotechnology research and almost all of the commercialization is being carried out by private firms based in industrialized countries.

This is a dramatic departure from the Green Revolution, in which the public sector played a strong role in research and technology diffusion. This paradigm shift has important implications for the kind of research that is performed, the types of technologies that are developed and the way these technologies are disseminated. The dominance of the private sector in agricultural biotechnology raises concerns that farmers in developing countries, particularly poor farmers, may not benefit - either because appropriate innovations are not available or are too expensive. Public-sector research was responsible for creating the high-yielding varieties of wheat and rice that launched the Green Revolution.

International and national public-sector researchers bred dwarfing genes into elite wheat and rice cultivars, causing them to produce more grain and have shorter stems and enabling them to respond to higher levels of fertilizer and water. These semi-dwarf cultivars were made freely available to plant breeders from developing countries who further adapted them to meet local production conditions. Private firms were involved in the development and commercialization of locally adapted varieties in some countries, but the improved germplasm was provided by the public sector and disseminated freely as a public good.

The countries that were able to make the most of the opportunities presented by the Green Revolution were those that had, or quickly developed, strong national capacity in agricultural research. Researchers in these countries were able to make the necessary local adaptations to ensure that the improved varieties suited the needs of their farmers and consumers.

National agricultural research capacity was a critical determinant of the availability and accessibility of Green Revolution agricultural technologies, and this remains true today for new biotechnologies. National research capacity increases the ability of a country to import and adapt agricultural technologies developed elsewhere, to develop applications that address local needs (*e.g.* “orphan crops”) and to regulate new technologies appropriately.

The biotechnology revolution, by contrast, is being driven largely by the private sector. Public-sector research has contributed to the basic science underpinning agricultural biotechnology, but the private sector is responsible for most applied research and almost all commercial development. Three interrelated forces are transforming the system for providing improved agricultural technologies to the world’s farmers.

The first is the strengthening environment for protecting intellectual property in plant innovations. The second is the rapid pace of discovery and the growing importance of molecular biology and genetic engineering. Finally, agricultural input and output trade is becoming more open in nearly all countries, enlarging the potential market for new technologies and older related technologies. These developments have created powerful new incentives for private research, and are altering the structure of the public/ private agricultural research endeavour, particularly with respect to crop improvemen.

With the growing importance of the private transnational sector, developing countries are facing increasing transaction costs in access to and use of technologies. Existing public-sector international networks for sharing technologies across countries and thereby maximizing spillover benefits are becoming increasingly threatened. The urgent need today is for a system of technology flows that preserves the incentives for private-sector innovation while at the same time meeting the needs of poor farmers in the developing world.

The organization and impacts of agricultural research and technology flows in the period 1960-90, when the Green Revolution paradigm of international, public-sector research held sway. The movement towards the increased privatization of agricultural research and development and its consequences for developing country access to technologies as revealed in recent global trends in biotechnology research, development and commercialization. A number of questions regarding the potential of the Gene Revolution to benefit the poor. These questions are taken up in the subsequent chapters of the report.

The world population may reach 10 billion by the middle of this century. Over the next 20 years, world cereal demand will increase by 50 per cent, driven by rapidly growing animal feed use and meat consumption. With the exception of acid-soil areas in Africa and South America, the potential for expanding global crop area is limited. Future expansions in food output must come largely from land already in use. The productivity of this land must be sustained and improved.

Most of the world's 842 million hungry people live in marginal lands and depend upon agriculture for their livelihoods. Food-insecure households in these higher-risk rural areas face frequent droughts, degraded lands, remoteness from markets and poor market institutions. For many of these people, food security will only come through increased agricultural production and income. Investments in science, infrastructure and resource conservation are needed to increase productivity and lower risks in marginal lands. Some of the problems in such environments will be too formidable to overcome. However, significant improvements should be possible. Biotechnology will play an important role in developing new germplasm with greater tolerance to abiotic and biotic stresses and with higher nutritional content. Continued genetic improvement of food crops - using conventional research tools and biotechnology - is needed to shift the yield frontier higher and to increase stability of yield.

Neolithic man - or much more likely woman - domesticated virtually all of our food and livestock species over a relatively short period, 10 000-15 000 years ago. Subsequently, several hundred generations of farmers were responsible for making enormous genetic modifications in all of our major crop and animal species. Thanks to the development of science over the past 150 years, we now have the insights into plant genetics and breeding to do purposefully what Nature did in the past by chance or design. Genetic modification of crops is not some kind of witchcraft; rather, it is the progressive harnessing of the forces of nature to the benefit of feeding the human race. Indeed, genetic engineering - plant breeding at the molecular level - is just another step in humankind's deepening scientific journey into living genomes. It is not a replacement for conventional breeding but a complementary research tool to identify desirable traits from remotely related taxonomic groups and transfer them more quickly and precisely into high-yielding, high-quality crop species. The world has the technology - already available or well advanced in the research pipeline - to feed on a sustainable basis a population of 10 billion people. However, access to such technology is not assured. The range of potential barriers includes issues related to intellectual property rights, technology acceptance by civil society and governments, and financial and educational barriers that keep poor farmers marginalized and unable to adopt new technology.

THE GREEN REVOLUTION: RESEARCH, DEVELOPMENT, ACCESS AND IMPACT

The Green Revolution was responsible for an extraordinary period of growth in food crop productivity in the developing world over the last 40 years. A combination of high rates of investment in crop research, infrastructure and market development, and appropriate policy support fuelled this progress. These elements of the Green Revolution strategy improved productivity growth despite increasing land scarcity and high land values.

Public-sector Research and International Technology Transfer

The Green Revolution defied the conventional wisdom that agricultural technology does not travel well because it is either agroclimatically specific, as in the case of biological technology, or sensitive to relative factor prices, as with mechanical technology. The Green Revolution strategy for food-crop productivity growth was explicitly based on the premise that, given appropriate institutional mechanisms, technology spillovers across political and agroclimatic boundaries could be created. Hence the Consultative Group on International Agricultural Research (CGIAR) was established specifically to generate technology spillovers, particularly for countries that are unable to capture all the benefits of their research investments. What happens to the spillover benefits from agricultural research and development in an increasingly global integration of food supply systems?

The major breakthroughs in yield potential that kick-started the Green Revolution in the late 1960s came from conventional plant-breeding approaches that initially focused on raising yield potential for the major cereal crops. The yield potential for the major cereals has continued to rise at a steady rate after the initial dramatic shifts in the 1960s for rice and wheat. For example, yield potential in irrigated wheat has been rising at the rate of 1 per cent per year over the past three decades, an increase of around 100 kg/ha/year. Essentially, no research or elite germplasm was available for many of the crops grown by poor farmers in less favourable agro-ecological zones (such as sorghum, millet, barley, cassava and pulses) during the early decades of the Green Revolution, but since the 1980s modern varieties have been developed for these crops and their yield potential has risen. In addition to their work on shifting the yield frontier of cereal crops, plant breeders continue to have successes in the less glamorous but no less important areas of applied research. These include development of plants with durable resistance to a wide spectrum of insects and diseases, plants that are better able to tolerate a variety of physical stresses, crops that require a significantly lower number of days of cultivation, and cereal grain with enhanced taste and nutritional qualities.

Prior to 1960, there was no formal system in place that provided plant breeders with access to germplasm available beyond their borders. Since then, the international public sector (the CGIAR system) has been the predominant source of supply of improved germplasm developed from conventional breeding approaches, especially for self-pollinating crops such as rice and wheat and for open-pollinated maize. These CGIAR-managed networks evolved in the 1970s and 1980s, when financial resources for public agricultural research were expanding and plant intellectual property laws were weak or non-existent. The exchange of germplasm is based on a system of informal exchange among plant breeders that is generally open and without charge. Breeders can contribute any of their material to the nursery and take pride in its adoption elsewhere in the world, while at the same time they are free to pick material from the trials for their own use.The international flow of germplasm has had a large impact on the speed and the cost of crop development programmes of national agricultural research systems (NARS), thereby generating enormous efficiency gains. Traxler and Pingali argued that the existence of a free and uninhibited system of germplasm exchange that attracts the best of international materials allows countries to make strategic decisions on the extent to which they need to invest in plant-breeding capacity. Even NARS with advanced crop research programmes, such as in Brazil, China and India, rely heavily on cultivars taken from these nurseries for their prebreeding material and for finished varieties. Small countries behaving rationally choose to free-ride on the international system rather than invest in large crop-breeding infrastructure of their own.

Evenson and Gollin report that, even in the 1990s, the CGIAR content of modern varieties was high for most food crops; 35 per cent of all varietal releases were based on CGIAR crosses, and an additional 22 per cent had a CGIAR-crossed parent or other ancestor. Evenson and Gollin suggest that germplasm contributions from international centres enabled developing countries to capture the spillover benefits of investments in crop improvement made outside their borders and achieve productivity gains that would have been more costly or even impossible had they been forced to work only with the genetic resources that were available at the beginning of the period.

IMPACTS OF FOOD-CROP IMPROVEMENT TECHNOLOGY

Substantial empirical evidence exists on the production, productivity, income and human welfare impacts of modern agricultural science and the international flow of modern varieties of food crops. Evenson and Gollin provide detailed information on the extent of adoption and impact of modern variety use for all the major food crops. The adoption of modern varieties (averaged across all crops) increased rapidly during the two decades of the Green Revolution, and even more rapidly in the following decades, from 9 per cent in 1970 to 29 per cent in 1980, 46 per cent in 1990 and 63 per cent by 1998. Moreover, in many areas and in many crops, first-generation modern varieties have been replaced by second- and third-generation modern varieties.

Much of the increase in agricultural output over the past 40 years has come from an increase in yield per hectare rather than an expansion of area under cultivation. For instance, FAO data indicate that for all developing countries, wheat yields rose 208 per cent from 1960 to 2000; rice yields rose 109 per cent; maize yields rose 157 per cent; potato yields rose 78 per cent and cassava yields rose 36 per cent. Trends in total factor productivity are consistent with partial productivity measures, such as rate of yield growth.

The returns to investments in high-yielding modern germplasm have been measured in great detail by several economists over the last few decades. Several recent reports have reviewed and analysed the data from hundreds of studies conducted over the last 30 years that calculated the social rates of return to investments in agricultural research. These studies examined investments by national and international public-sector institutions in Africa, Asia, Latin America and the Organisation for Economic Co-operation and Development (OECD) countries as well as by the private sector. Although these studies were carried out using a variety of different methods, they showed considerable consistency. The average social rate of return to public investment in agricultural research reported in these studies is in the region of 40-50 per cent. Private-sector research was also found to generate similar rates of social returns.

The primary effect of agricultural research on the non-farm poor, as well as on the rural poor who are net purchasers of food, is through lower food prices. The wide-spread adoption of modern seed-fertilizer technology led to a significant shift in the food supply function, increasing output and contributing to a fall in real food prices:

- The effect of agricultural research on improving the purchasing power of the poor - both by raising their incomes and by lowering the prices of staple food products - is probably the major source of nutritional gains associated with agricultural research. Only the poor go hungry. Because a relatively high proportion of any income gains made by the poor is spent on food, the income effects of research-induced supply shifts can have major nutritional implications, particularly if those shifts result from technologies aimed at the poorest producers.

Studies by economists have provided empirical support for the proposition that growth in the agriculture sector has economy-wide effects. Hayami *et al.* illustrated at the village level that rapid growth in rice production stimulated demand and prices for land, labour and non-agricultural goods and services. For sector-level validation of the proposition that agriculture does indeed act as an engine of overall economic growth.

Once modern varieties have been adopted, the next set of technologies that makes a significant difference in reducing production costs includes machinery, land management practices (often in association with herbicide use), fertilizer use, integrated pest management and (most recently) improved water management practices.

Although many Green Revolution technologies were developed and extended in package form (*e.g.* new plant varieties plus recommended fertilizer, pesticide and herbicide rates, along with water control measures), many components of these technologies were taken up in a piecemeal, often stepwise manner. The sequence of adoption is determined by factor scarcities and the potential cost savings achieved. Herdt provided a detailed assessment of the sequential adoption of crop management technologies for rice in the Philippines. Traxler and Byerlee provided similar evidence on the sequential adoption of crop management technologies for wheat in Sonora, northwestern Mexico. Although the favourable, high-potential environments gained the most from the Green Revolution in terms of productivity growth, the less favourable environments also benefited through technology spillovers and through labour migration to more productive environments. According to David and Otsuka (1994), wage equalization across favourable and unfavourable environments was one of the primary means of redistributing the gains of technological change. Renkow found similar results for wheat grown in high- and low-potential environments in Pakistan. Byerlee and Moya, in their global assessment of the adoption of modern varieties of wheat, found that over time the adoption of modern varieties in unfavourable environments caught up with those in more favourable environments, particularly when germplasm developed for high-potential environments was further adapted to the more marginal environments. In the case of wheat, the rate of growth in yield potential in drought-prone environments was around 2.5 per cent per year during the 1980s and 1990s. Initially, the growth in yield potential for the marginal environments came from technological spillovers as varieties bred for the high-potential environments were adapted to the marginal environments. During the 1990s, however, further gains in yield potential came from breeding efforts targeted specifically at the marginal environments. Even while highlighting the yield breakthrough in wheat, the Government had also launched a massive programme to develop and spread high-yielding varieties for rice, maize, sorghum and pearl millet. These programmes were the drivers of the "Green Revolution" in India, which permitted striking advances in production and productivity without increasing cultivated area.Because these high-yielding varieties require inputs such as fertilizer and irrigation water, social scientists criticized the Green Revolution technologies for not being resource neutral. Environmentalists attacked the Green Revolution because of potential damage to long-term productivity as a result of excessive use of pesticides and fertilizers and monocropping. Despite the success of the Green Revolution in raising millions of people out of misery, the incidence of poverty, endemic hunger, communicable diseases, infant and maternal mortality rates, low birth-weight children, stunting and illiteracy remain high.

The concerns of social scientists and ecologists and the remaining urgent problems of poverty and hunger led to my developing the concept of an "evergreen revolution" to stress the need for enhancing crop productivity in perpetuity without associated ecological or social harm. An evergreen revolution can be achieved only if we pay attention to pathways that can help to achieve revolutionary progress in enhancing productivity, quality and value-addition under conditions of diminishing per capita arable land and irrigation water availability, expanding biotic and abiotic stresses, and fast-changing consumer and market preferences. This will require mobilizing the best in both traditional wisdom and technologies and frontier science. Among the frontier technologies relevant to the next stage in our agricultural revolution, the foremost is biotechnology.

The apprehensions relating to molecular genetics and genetic engineering fall under the following broad categories: the science itself, the control of the science, access to the science, environmental concerns, and human and animal health. A disaggregated approach to the study of these issues will be important for a rigorous analysis of risks and benefits. Dealing with these issues in a composite manner for all applications of genetic engineering will result in inappropriately broad conclusions, such as the general condemnation of GMOs expressed by non-governmental organizations (NGOs) at the World Food Summit: *five years later* held in Rome in 2002.

The benefits of molecular breeding techniques such as the use of molecular markers and undertaking precision breeding for specific characters through recombinant DNA technology are immense. The work already performed in India has revealed the potential for breeding new GM varieties possessing tolerance to salinity, drought and some major pests and diseases, together with improved nutritive quality. A new era of integrated Mendelian and molecular breeding has begun.

An evergreen revolution will blend these frontier technologies with the ecological prudence of traditional communities to create technologies that are based on integrated natural resource management and that are location specific because they are developed through participatory experimentation with farm families.

This is the only way we can face the challenges of the future, particularly in the context of the growing water scarcity and the urgent need to step up productivity in semi-arid and dry farming areas. Accelerated agricultural progress is the best safety net against hunger and poverty, because in most developing countries over 70 per cent of the population depend on agriculture for their livelihood. Denying ourselves the power of the new genetics will be doing a great disservice both to resource-poor farming families and to the building of a sustainable national food and nutrition system.

BIOTECHNOLOGY IN FOOD AND AGRICULTURE

Biotechnology in food and agriculture, particularly genetic engineering, has become the focus of a "global war of rhetoric". Supporters hail genetic engineering as essential to addressing food insecurity and malnutrition in developing countries and accuse opponents of "crimes against humanity" for delaying the regulatory approval of potentially life-saving innovations. Opponents claim that genetic engineering will wreak environmental catastrophe, worsen poverty and hunger, and lead to a corporate takeover of traditional agriculture and the global food supply. They accuse biotechnology supporters of "fooling the world" of *The State of Food and Agriculture* surveys the current state of scientific and economic evidence regarding the potential of agricultural biotechnology, particularly genetic engineering, to meet the needs of the poor.

Agriculture in the twenty-first century is facing unprecedented challenges. An additional 2 billion people will have to be fed over the next 30 years from an increasingly fragile natural resource base. More than 842 million people are chronically hungry, most of them in rural areas of poor countries, and billions suffer from micronutrient deficiencies, an insidious form of malnutrition caused by the poor quality of, and lack of diversity in, their habitual diet. The Green Revolution taught us that technological innovation - higher-yielding seeds and the inputs required to make them grow - can bring enormous benefits to poor people through enhanced efficiency, higher incomes and lower food prices. This virtuous cycle of rising productivity, improving living standards and sustainable economic growth has lifted millions of people out of poverty. But many remain trapped in subsistence agriculture. Can the Gene Revolution reach those left behind?

At the same time, a rapidly urbanizing global population is demanding a wider range of quality attributes from agriculture, not just of the products themselves but of the methods used in their production. The agriculture sector will need to respond in ways beyond the traditional focus on higher yields, addressing the protection of environmental common goods, consumer concerns for food safety and quality, and the enhancement of rural livelihoods both in the South and in the North. Is the rhetoric of war deafening us to a more reasoned debate regarding the hazards and opportunities posed by biotechnology?There is clear promise that biotechnology can contribute to meeting these challenges. Biotechnology can overcome production constraints that are more difficult or intractable with conventional breeding. It can speed up conventional breeding programmes and provide farmers with disease-free planting materials. It can create crops that resist pests and diseases, replacing toxic chemicals that harm the environment and human health, and it can provide diagnostic tools and vaccines that help control devastating animal diseases. It can improve the nutritional quality of staple foods such as rice and cassava and create new products for health and industrial uses.

But biotechnology is not a panacea. It cannot overcome the gaps in infrastructure, markets, breeding capacity, input delivery systems and extension services that hinder all efforts to promote agricultural growth in poor, remote areas. Some of these challenges may be more difficult for biotechnology than for other agricultural technologies, but others may be less difficult. Technologies that are embodied in a seed, such as transgenic insect resistance, may be easier for small-scale, resource-poor farmers to use than more complicated crop technologies that require other inputs or complex management strategies. On the other hand, some biotechnology packages, particularly in the livestock and fisheries areas, require a certain institutional and managerial environment to function properly and thus may not be effective for resource-poor smallholders.

The safety and regulatory concerns associated with transgenic crops constitute a major hurdle for developing countries, because many lack the regulatory frameworks and technical capacity necessary to evaluate these crops and the conflicting claims surrounding them. Although the international scientific community has determined that foods derived from the transgenic crops currently on the market are safe to eat, it also acknowledges that some of the emerging transformations involving multiple transgenes may require additional food-safety risk-analysis procedures. There is less scientific consensus on the environmental hazards associated with transgenic crops, although there is general agreement that these products should be evaluated against the hazards associated with conventional agriculture. There is also wide consensus that transgenic crops should be evaluated on a case-by-case basis, as is the case with pharmaceuticals, taking into consideration the specific crop, trait and agro-ecological system. Because very few transgenic crops have been evaluated for their ecological impacts in tropical regions, a major research effort is required in this area. Public- and private-sector transgenic crop research and development are being carried out on more than 40 crops worldwide and dozens of innovations are being studied, but there is clear evidence that the problems of the poor are being neglected. Barring a few initiatives here and there, there are no major public- or private-sector programmes to tackle the critical problems of the poor or targeting crops and animals that they rely on.

Concerted international efforts are required to ensure that the technology needs of the poor are addressed and that barriers to access are overcome.

DEVELOPMENT OF PLANT BIOTECHNOLOGY

Corn (maize) originated in Mexico from a grass called teosinte that has a small reproductive structure bearing little resemblance to the ear of corn seen in markets around the world today. Tomatoes and potatoes first appeared in South America - tomatoes as small fruits the size of a grape, and potatoes as knobby tubers with high concentrations of a family of bitter chemicals called glycoalkaloids, which are toxic to humans.

Through selective breeding by our ancestors, the shape, colour and chemical content of these and hundreds of other plants consumed today have been modified to suit consumer preference or to obtain desired characteristics such as high yield, disease and insect resistance, and tolerance to drought and other plant stresses. Not only have these plants changed in appearance and composition, they also have become distributed worldwide through centuries of human migration and trade. For example, cabbage, which originated in Europe, is now grown on every inhabited continent. When today's consumers walk into a market in many parts of the world, they are witnesses to today's global food system where foods produced in one part of the world are daily shipped to local markets.

We now realise our ancient ancestors were modifying the genetic makeup of plants by transferring genetic material from one plant to another. However, it wasn't until Gregor Mendel, an Austrian monk, conducted experiments in the 1800s with garden peas that the basic laws of heredity were first unraveled. Prior to the early 1900s, traditional plant breeding, like that practiced by Mendel, relied on man-made artificial crosses in which pollen from one plant species was transferred to another sexually compatible plant.

The goal was to take a desirable trait from one plant and introduce it into another plant. However, often desirable characteristics either were not present in sexually compatible plants or did not occur in any plant species. This led plant breeders to seek new ways of transferring desirable genes. Beginning in the 1930s, plant breeders developed techniques to allow them to develop plants from two parent plants that could not normally produce viable offspring. An example is the technique called "embryo rescue," in which the new plant embryo is provided with extra care in the laboratory to enable it to survive during its early growth.

In the 1950s, plant breeders also developed methods of creating variation in an organism's genetic structure through what is termed "mutation breeding." Mutations in the genetic makeup of a plant occur continuously and randomly in nature through such events as the sun's radiation and may lead to the occurrence of new desirable traits. Mutation breeding uses similar random processes to cause changes in a plant's genes. Plants then are assessed to determine if the genes were changed and whether the changes provided a beneficial trait such as disease or insect resistance. If the plant was "improved," then it was tested for other changes that may have occurred. Many of the common food crops we use daily have been developed through techniques such as embryo rescue and mutation breeding and virtually all the foods we consume have genes in them.

It is hard to think of an example of a common food crop in the developed world that has not been improved by some form of modern technology, or what can be termed "biotechnology." Simply put, biotechnology is a set of techniques that utilizes living organisms, or parts of organisms, to make or modify products, improve plants or animals, or develop microorganisms for specific purposes. This definition encompasses all human activities conducted on living organisms from the earliest development of plant breeding 10,000 years ago to the present. This is the reason plant breeders consider the term "genetically modified organisms" - or GMOs - to be misnomer since all common food crops of today have been so modified.

THE SCIENCE OF MODERN GENETIC ENGINEERING

Genetic engineering is one form of biotechnology and usually refers to copying a gene from one living organism—plant, animal or microbe—and adding it to another organism. In genetic engineering, a small piece of genetic material (DNA) is inserted into another organism to produce a desired effect. This is in contrast to traditional plant breeding in which all the genes desirable and undesirable contained in the male plant—pollen—are combined with all the genes of the female plant.

The progeny resulting from this cross may contain the gene for the desirable character, but it will also contain many of the undesirable genes from both parents. Genetic engineering has the advantage of being able to transfer only the gene of interest and greatly accelerate plant breeding. But genetic engineering also is more powerful than traditional breeding since it can move genes not only between similar plant species but also from distant relatives, including non-plant species. It is possible to move genes between such seemingly unrelated organisms because all living organisms share the same code for DNA and the synthesis of proteins and other basic life functions. What might seem on the surface to be very different organisms are, in fact, very similar, at least at the molecular level. All living things are more alike than different, and this is one of the reasons that genes can be moved so successfully between such seemingly different organisms as plants and bacteria. Genes are not unique to the organisms from which they came, so there really aren't "tomato genes" or "bacterial genes." It's the collection of all genes in a tomato or a bacterium that makes it a tomato or bacterium, not a single gene. As we learn more about the genetic makeup of all organisms, we see that most plant species differ by only a small percentage of their genes and those even such seemingly different organisms as tomatoes and bacteria have many common genes.

These findings suggest that in the long-term evolutionary process even tomatoes and bacteria had some common ancestor. From the discovery 50 years ago of the structure of DNA, scientists soon came to realise they could take segments of DNA that carried information for specific traits—genes—and move them into another organism. In 1972, the collaboration of Hubert Boyer and Stanley Cohen resulted in the first isolation and transfer of a gene from one organism to a single-celled bacterium where it would express the gene and manufacture a protein. Their discoveries led to the first direct use of biotechnology—the production of synthetic insulin to treat people with diabetes—and the start of what is often called modern biotechnology. Plants were first transformed through genetic engineering in the late 1970s. Mary-Dell Chilton and colleagues used a common soil-dwelling bacterium, Agrobacterium tumefaciens, that attaches itself to plants and transfers some of its DNA into the plant. Chilton and her colleagues added a gene to this bacterium, which in turn transferred the gene into a plant where it became part of the plant's DNA. This bacterium is still commonly used in genetic engineering along with another technique that uses a high-velocity mechanism to inject DNA into plant cells. The result from either technique is the same—the plant cells take up the gene and begin to express it as their own.

Benefits and Risks of Biotechnology

Plants developed through genetic engineering were first grown on 1.7 million hectares in 1996 in the United States, but by 2002 they were grown on 58.7 million hectares in 16 countries. By far the major use of the present plants is to manage pests—weeds, insects and diseases. Weed management with genetically engineered plants is accomplished because the plants have a modified enzyme (a protein) that allows them to survive an application of a specific herbicide that normally acts on that enzyme. Growers can plant the herbicide-tolerant seeds, allow the plants to emerge along with any weeds in the field and then treat the field with an herbicide.

The result is that the weeds, but not the crops, die. The advantage to growers is that they spend less time on weed management, have enhanced weed control, use safer herbicides, and in many cases use less herbicides. Additionally, this technology allows growers to use soil conservation practices such as reduced or no-tillage, thus helping to retain soil structure and moisture and reduce erosion. Herbicide tolerant crops (soybean, canola, cotton and maize) were grown on 48.6 million hectares in 2002. Insect-resistant crops developed through genetic engineering utilize the common soil bacterium, Bacillus thuringiensis (Bt), which has been commercially used for more than 50 years, as an insecticide spray.Although safe to humans and the environment, when a susceptible insect ingests Bt, the Bt protein binds to specific molecular receptors in the gut and creates pores causing the insect to starve to death. Insecticidal products containing Bt were first commercialized in France in the late 1930s, but even in 1999 the total sales of Bt products constituted less than 2 per cent of the total value of all insecticides. Bt, which had limited use as a foliar insecticide, became a major insecticide only when genes that produce Bt toxins were engineered into major crops. The Bt crops available at present are maize and cotton. These were grown on a total of 14.5 million hectares in 2002. Virus-resistant crops were created by inserting a non-infective part of a plant virus into a plant, essentially "vaccinating" the plant to protect it from the virus.

This technique is called "pathogen-derived resistance." Squash and papaya have been engineered to resist infection by some common viruses and are approved for sale in the United States. There are fewer than 1 million hectares of these crops. The bioengineered plants available at present provide growers with better tools to manage pest problems. As with any technology, there are risks and benefits to currently available genetically engineered plants, but the present body of information indicates their use has enhanced pest management, substantially reduced the amounts of pesticides used in some crops, enabled growers to use safer pesticides, and contributed to enhanced safety for humans and the environment.

The regulatory process for managing these plants and their effects on the environment and humans has evolved with the technology and the scientific community's knowledge of these tools. Many of the more controversial issues surrounding genetic engineering of plants—such as pesticide resistance, gene flow and intellectual property issues—are not unique to this new technology but pertain to all types of agriculture. Some species of insects have developed resistance to sprays of Bt, indicating the potential for some species to become resistant to Bt plants.

However, despite Bt plants being grown on more than 62 million hectares worldwide from 1996 to 2002, there have been no documented cases of resistance development. The reasons for this lack of resistance appear to involve not only biological factors of the insects and the Bt plant, but also the fact that the regulatory agency (the Environmental Protection Agency) in the United States requires a resistance management plan for growing Bt plants. No other insecticide has such strict regulations. Still, growers, companies and federal regulatory agencies must be vigilant about resistance developing for biotech crops used to manage insects, weeds and viruses as they also must with non-biotech pest management tactics.

It will be important to consider the accrued environmental and health benefits of these biotech crops prior to the development of any resistance and how resistance can be managed if and where it occurs. In addition to pesticide resistance, gene flow from biotech to non-biotech crops may also be a concern. However, the risk of gene flow varies with each crop and each gene. Pollen flow in soybeans is very limited so the risk of a biotech soybean crop crossing with a non-biotech soybean is minimal, but this may be different for another crop. Likewise, if the gene in the biotech crop that provided a pest management trait, such as insect resistance, moved into a non-biotech plant, such as a weed, any selective advantage of the insect-protected weed in the ecosystem should be assessed. These same questions should also be answered with non-biotech crops, but these have not received the same level of attention as biotech crops because of the latter's higher profile.

Horizon of Biotechnology

In the future, the potential uses of plant biotechnology are far more wide-ranging than the pest-management biotech crops of today. Plants are being developed that serve as production "factories" for medically important drugs, sources of alternative energy, tools for cleaning toxic waste sites, and biomaterials including dyes, inks, detergents, adhesives, lubricants, plastics and the like. Consumers may see these products as more directly enhancing their quality of life than the pest-management biotech crops of today. Perhaps an even more dramatic advantage to consumers will be seen when plants are genetically engineered to have enhanced health benefits such as disease-fighting chemicals or increased amounts of essential vitamins and minerals. A healthy and well-informed discussion of the risks and benefits involved in agricultural biotechnology is needed to ensure a proper role for this new technology in our future food and health systems. No one should believe that any technology, including biotechnology, will completely solve the world's agricultural problems. Many people familiar with biotechnology, however, believe it to be an important component of the solution.

IMPROVING ANIMAL AGRICULTURETHROUGH BIOTECHNOLOGY

Livestock feed derived from biotechnology has been shown to increase production efficiency, decrease animal waste and lower the toxins that can cause sickness in animals, asserts Terry Etherton, distinguished professor at The Pennsylvania State University. Genetically modified feed also can improve water and soil quality by reducing levels of phosphorous and nitrogen in animal waste. Over the past 20 years, biotechnology has lead to the development of new processes and products that have benefited agriculture and society.

Between 1996 and 2002 there was a 35-fold increase in acreage planted globally with genetically modified (GM) crops, from 1.7 to 58.1 million hectares, and more than a quarter of GM crops are grown in developing countries. While there has been considerable discussion about the benefits of GM crops in the grains and fruits humans consume, less public debate has been forthcoming about GM crops' profound effects on improving the health of livestock grown for meat products and on reducing some of the environmental costs of livestock wastes. Adoption of products produced by modern biotechnology will be important to enable the production of food sufficient to feed a growing world population. Biotechnologies that enhance productivity and productive efficiency—feed consumed per unit of milk or meat produced—have been developed and approved for commercial use in many countries. New biotechnology products have enabled improvements to be made that increase food safety and improve animal health. Biotechnology also offers considerable potential to animal agriculture as a means to reduce nutrients and odours from manure and volume of manure produced. Development and adoption of these biotechnologies will contribute to a more sustainable environment.

In order to be approved for commercial use in the United States, new agricultural biotechnologies are evaluated rigorously by the appropriate federal regulatory agencies to ensure efficacy, consumer safety, and animal health and well-being. Successful development and adoption of emerging biotechnologies for agriculture require improved public understanding of scientific, economic, legislative, ethical and social issues. The objective of this chapter is to provide a brief overview of some of the existing and emerging modern agricultural biotechnologies that affect animal productivity and discuss their current or potential food safety and environmental benefits.

FEEDING LIVESTOCK

Scientific studies evaluating feed components derived from genetically modified (GM) plants have focused on beef cattle, swine, sheep, fish, lactating dairy cows, and broiler and layer chickens, and have included nutrient composition assessments, digestibility determinations and animal performance measurements. These studies have shown that feed components derived from GM plants are equivalent in terms of nutrient composition to non-GM plants.

Feeding components derived from GM plants, such as grain, silage and hay, also show results in growth rates and milk yields that are equivalent to those food components derived from non-genetically enhanced feed sources. Studies have reported that GM maize altered for protection against the "corn borer" can have lower contamination by mycotoxins—toxic substances produced by fungi or molds—under certain growing conditions, resulting in safer feed for livestock. In the United States, there is a long history of assessing the safety of foods introduced into the marketplace. The assessment of GM plants and animal biotechnologies is science-based and rigorous. The discovery and development of ne

In the United States, there is a long history of assessing the safety of foods introduced into the marketplace. The assessment of GM plants and animal biotechnologies is science-based and rigorous. The discovery and development of new animal and plant biotechnologies are part of a continuum leading to the commercialization of agricultural biotechnology products. Historically, equivalence of composition GM plants, GM animals or animals treated with biotechnology products, such as bST, has been an important component of the regulatory process. Establishing equivalence of composition is evidence that substantive changes did not occur in the plant or animal as the result of the genetic modification event.

One endorsement of the robust nature of the comparative safety assessment process used with GM plants is that more than 223 million hectares of GM crops have been commercially grown over the past 10 years with no documented effects to humans, animals or the environment. Likewise, there have been no documented adverse effects of meat and milk derived from cows supplemented with bST, the most rapidly adopted animal biotechnology to date.

Agriculture is transiting a remarkable scientific era with respect to the myriad of processes and products that have been developed using biotechnology. Moreover, many new products of biotechnology are being developed that will benefit the food sector.

Implicit to approval of these new products is a robust safety assessment process. To date, the approved GM plants and animal biotechnologies have been judged to be as safe as conventionally produced counterparts. Development and adoption of new biotechnologies will be crucial in meeting the challenge of producing enough food for a growing world population while reducing impacts on the environment. The impact these technologies have on society in the future, however, will be largely dependent on the extent to which they are adopted by producers and the agricultural community and accepted by consumers. Questions about societal impacts and safety often arise as the result of technological change. Inherent to the successful development and adoption of new biotechnologies for agriculture is the need to increase public understanding of the scientific, economic, legislative, ethical and social issues associated with emerging agricultural biotechnologies.

THE IMPACTS OF AGRICULTURALPRACTICES ON BIODIVERSITY

The impacts of common agricultural practices on biodiversity, and ways in which some of these impacts can be mitigated. The biodiversity can be quantified in several different, equally important ways, thus agricultural impacts on biodiversity are considered both in terms of species and genetic diversity.

Within each of these categories, the impacts on agricultural biodiversity and natural biodiversity are addressed separately because the impacts of agriculture are different on these two types of habitat. This distinction could also be thought of as on-site and off-site impacts of agricultural practices.

IMPACTS ON SPECIES BIODIVERSITY

Modern agricultural practices have been broadly linked to declines in biodiversity in agro-ecosystems. This has been found to be true for a wide variety of taxonomic groups, geographic regions and spatial scales. More specifically, various researchers have found significant correlations between reductions in biodiversity at various taxonomic levels and measures of agricultural intensification.

For example, a review of published studies on arthropod diversity in agricultural landscapes found species biodiversity to be higher in less intensely cultivated habitats. Similarly, analysis of 30 years of monitoring records demonstrated that the abundance of aerial invertebrates at a location in rural Scotland was negatively correlated with a suite of agricultural variables that represent more intensive agriculture; that is, arthropod populations are lowest where agriculture is the most intensive.

In this same study, the abundance of various farmland bird species was, in turn, positively correlated with arthopod abundance in the same year and the previous year. Comparable studies have found similar impacts on bird species throughout the United Kingdom and European Union (EU). Across Europe, declines in farmland bird diversity are correlated with agricultural intensity and declines in the European Union have been greater than in non-Member States. These effects of agricultural intensification undoubtedly reflect a large number of factors which are addressed individually in the following sections, including the cropping pattern, the frequency of tillage, the amount and nature of fertilizers used, and the amount and nature of pesticides applied (particularly insecticides and herbicides). However, it should be kept in mind that all of these factors are interrelated to a greater or lesser degree.

Crop Diversity

Intensive agricultural systems typically have limited crop diversity. Many such systems are monocultures at least at the level of individual fields, and are relatively homogenous even at the regional level.

Low crop diversity generally will mean both limited botanical diversity and limited structural diversity. Robinson and Sutherland analysed changes in agriculture and biodiversity in Britain since the 1940s. They found a consistent reduction in landscape diversity, as reflected in a 65% decline in the number of farms. Farms had become more specialized and efficient. This also was associated with the removal of 50% of hedgerows and a reduction in winter stubbles. Hedgerows and similar non-cropped habitat are important sources of food and shelter for a variety of plants and invertebrates.Reductions in landscape diversity lead to lower faunal diversity in intensively managed agro-ecosystems than in more diverse agricultural systems or in natural habitats. For example, Robinson and Sutherland found major declines in organisms associated with farmland in Britain and northwest Europe, particularly in habitat specialists. As an illustration, biodiversity declines in bird species were related to reduced food availability in the non-breeding season.

They concluded that reduced habitat diversity was of particular important in the 1950s and 1960s, while reduction in habitat quality may be more important now. Similarly, a review of the available literature on arthropod diversity found that structural biodiversity in agricultural areas is correlated with functional and species biodiversity of the above-ground insect fauna.

Tillage

Intensive tillage leads to frequent disturbances of the agricultural landscape, increases energy loss from agricultural fields, and increases problems of soil erosion and run-off from agricultural fields. All of these factors adversely affect the quality of agricultural habitats, with significant consequences for agricultural biodiversity. When looked at corn, soybean and wheat cropping systems in the Mid- Atlantic region of the United States, they found that ground-dwelling and foliage-dwelling beneficial arthropods were least abundant, and pests were most abundant, in the simplest, most intensively managed continuous corn system.

In general, ground-dwelling species were more abundant in no-till than in deep-tilled crops. This suggests that shifts towards conservation tillage and no-till will benefit agricultural biodiversity. Such shifts have been occurring recently in many cropping systems as farmers recognize the environmental and economic benefits of doing so.

Pesticide Use

Adverse effects of pesticide use in agriculture are well-documented. Conventional insecticides generally reduce diversity through direct toxic effects. Many of the widely used classes of conventional insecticides, including organophosphates and pyrethroids, have been shown to adversely affect a broad range of non-target species, including species of economic importance. Local extinctions are common where these insecticides are frequently used. Such insecticides have been shown to eliminate important predator and parasitoid species from agricultural systems. In Indian cotton for example, over 600 such species have disappeared altogether. These impacts on natural enemies have been shown to lead to flare-ups in secondary pest species, some of which were not previously economically important. In a few cases, insecticides directly stimulate the population growth of non-target pest species, *e.g.* pyrethroids have such an effect on some mite and aphid species. In addition, the toxic effects of insecticides can lead to food chains effects because of decreased food availability for higher trophic levels and bioaccumulation of the insecticides. For example, organochlorine use and ingestion by earthworms has led to die-offs of birds feeding on these species. Replacing broad-spectrum insecticides with more specific, softer alternatives is necessary to avoid these impacts. Some herbicides also can be toxic to invertebrates. However, the more important effects of herbicide use with respect to biodiversity are to reduce non-crop plant (weed) populations and weed seed production in agricultural fields. Where herbicide use is intensive, adverse impacts may be seen on various vertebrate and invertebrate species that depend upon these plant species for food or shelter. Where invertebrate populations are strongly affected, consequences for higher trophic levels also may occur.

Genetically Modified (GM) Crops

The use of GM crops can positively impact agricultural species biodiversity. In particular, the adoption of insect resistant Bt crops, expressing highly specific Bt proteins, represents an opportunity to replace broad-spectrum insecticide use. The insecticidal proteins expressed in Bt crops such as Bt maize and Bt cotton are so narrow in their activity that they have little or no activity against non-target organisms. Furthermore, the toxins are expressed within the plant tissues, minimizing the exposure to animals that do not feed on the crop plants.

As a consequence, across the large number of field studies that have been conducted, few or no differences have been seen with respect to community structure or individual species abundances where fields of Bt crops have been compared to conventional crops that have not been treated with insecticides. Where they have been calculated, indices of species diversity and community structure have not differed significantly for Bt corn fields compared to untreated conventional corn fields or for Bt cotton fields compared to conventional cotton fields. The only species that have been observed to be significantly and consistently less abundant in fields of Bt crops relative to fields of conventional crops are the target pests.

In studies where the conventional crop fields have been sprayed for the target pest species of the Bt crop (as rountinely occurs in most crop systems), many non-target species have been observed to be adversely impacted, leading to significantly lower non-target populations in sprayed conventional fields as compared to Bt crop fields.

With corn fields, this is particularly obvious for foliage-dwelling species because of the method of application of these insecticides, but ground-dwelling species like carabids and cursorial spiders are also often affected, directly or indirectly, by the insecticidal sprays and are apparently not affected by Bt corn. Similarly, a variety of studies of Bt cotton in the United States, Australia and China have all demonstrated that populations of many non-target species are higher in Bt cotton fields than in sprayed conventional cotton fields. Likewise, work on potato fields in the northeastern US has revealed larger populations of many generalist predators in Bt potato fields than in conventional potato fields treated with appropriate broad-spectrum insecticides.

The years long controversy on the fate of the monarch larvae in the US cornfields seems to be solved: After the first shock of the Nature publication of extensive field work demonstrated no significant impact. It was Rachel Carson herself who named Bt proteins as a possible way out of the pesticide crisis which she described in her famous 'The Silent Spring', and one can only wonder what she would have said about the Bt toxin instead of being sprayed in large, but rapidly decomposing quantities built genetically into the corn borer infested crops.

Herbicide tolerant crops are not expected to directly affect agricultural biodiversity because of the nature of the proteins expressed but they may lead to changes in practices that could affect biodiversity. Herbicide tolerant crops facilitate shifts towards reduced tillage, as observed for soybean and cotton in the United States. As noted earlier, such shifts can be beneficial to agricultural eco-systems.

In addition, herbicide tolerant crops permit greater flexibility in herbicide application practices, particularly the timing of applications. If these practices lead to more intensive and higher level weed control, then biodiversity may be adversely affected. However, herbicide tolerant crops also can encourage herbicide application practices that benefit wildlife.

For example, studies on herbicide tolerant sugarbeet in the UK and Denmark have shown that leaving weeds untreated in the agricultural field for a longer period allow arthropod populations to increase to higher levels than are seen in conventional fields, without affecting crop yield. These weeds and the associated arthropods provide valuable food and habitat for farmland birds and other wildlife. Such a practice is not feasible with conventional sugarbeets. Soil fertility can be enhanced with appropriate use of broad spectrum herbicide tolerant sugar beets.

The Fate of Bt Toxin in the Soil

It has been shown that Bt toxin is released into the rhizosphere soil with decaying litter and through root exudates from Bt corn. The insecticidal toxin produced by B. thuringiensis subsp. kurstaki remains active in the soil, where it binds rapidly and tightly to clays and humic acids. The bound toxin retains its insecticidal properties as determined by bioassays: the toxin is protected for some time against microbial degradation by being bound to soilparticles, persisting in various soils for at least 234 days (the longest time studied).Unlike the bacterium, which produces the toxin in a precursor form, Bt corn contains an inserted truncated cry1Ab gene that encodes the active toxin. The toxins do not appear to have any consistent effects on organisms in soil (earthworms, nematodes, protozoa, bacteria, fungi) or on microorganisms *in vitro*. A multiseason monitoring in six fields in the USA did not reveal any effect on various bioassays with soil organisms, using soil matter including degrading leafes. A recent study focussing on bioassays with degrading leaf litter of two near isolines of Bt- and non-Bt-maize under controlled conditions. The study concludes that possible subtle, longterm toxic effects should be tested in long term monitoring in the post-commercialization phase. These possible effects should be put into quantitative relation to long term monitoring data under field conditions with non-Bt maize, where more pesticides are used.

Agricultural practices adversely affect in-field biodiversity in a number of obvious ways.

Most of these practices can be effectively mitigated through judicious use of available technologies and crop management strategies. For example, GM crops can replace agricultural practices that would otherwise depress and disrupt species biodiversity, and can encourage or complement other practices that enhance biodiversity. Existing agricultural policies and other political factors also strongly affect the decisions made by farmers. That said, the potential exists to directly or indirectly reward farmers for making environmental improvements to their land.

NATURAL BIODIVERSITY

General Impacts of Modern Intensive Agriculture

The experts generally agree that the factor most responsible for decreases in natural biodiversity, both locally and globally, is habitat destruction and fragmentation as land with native communities is cleared for agricultural or other use. The past 35 years have brought a 1.68-fold increase in the amount of irrigated cropland and a 1.1-fold increase in cultivated land.

This problem is most severe in developing countries with a large amount of subsistence agriculture. Even looking within different agricultural systems, the degree of fragmentation increases with the intensity of agricultural management. For example, Belanger and Grenier showed that fragmentation increased along a gradient from traditional dairy agriculture to more intensive cash crop agriculture in the St. Lawrence Valley of Quebec, Canada.

Pesticide Use

Pesticide use, and particularly insecticide use, has significant off-site effects on biodiversity. Aerial drift and movement in water can expose natural communities to potentially toxic amounts of pesticides. These effects will be most severe on communities adjacent to agricultural lands. Boutin and Jobin found the species composition in habitats adjacent to agricultural habitats to be adversely affected by intensive agriculture, as measured by tillage practices and pesticide and fertilizer use. These non-crop habitats adjacent to cropped land are critical for the maintenance of plant species diversity, for the conservation of beneficial pollinating and predatory insects, and as essential habitat for birds.

Tillage and Fertilizer Use

The impacts of tillage on biodiversity in agricultural fields were described earlier, the disruption of in-field communities and reduction of soil quality being the most obvious. However, the impacts of tillage on natural habitas are even greater. Soil erosion due to tillage leads to high levels of fertilizers and pesticides being carried off agricultural fields into waterways. Remember that the past 35 years have seen a 6.87-fold increase in nitrogen fertilization and a 3.48-fold increase in phosphorus fertilization within intensive agricultural systems.

As they move into aquatic systems, these chemicals can have direct toxic effects on natural communities, while the fertilizers cause eutrophication. Eutrophication leads, in turn, to direct losses in biodiversity, pest outbreaks, and changes in the structure of natural communities. In addition, because erosion leads to various forms of nitrogen and particultae matter being redistributed aerially, natural terrestrial ecosystems also are being eutrophied. Many of these problems can be reduced or avoided by reducing tillage practices. In North America and Europe, high-yield farming and conservation tillage have reduced soil erosion by 65-98%. However, subsistence farming in developing countries is causing substantial soil erosion and habitat loss, and is a significant threat to natural biodiversity.

Genetically Modified Crops

GM crops have the ability to benefit natural biodiversity in a number of ways. *First*, GM crops have the demonstrated potential to increase yields and decrease variability in yields thereby reducing the need to put additional land into agricultural production. By slowing the rate at which natural habitats are destroyed, GM crops and other technologies that increase agricultural productivity can help to preserve natural biodiversity. *Second*, insect resistant crops reduce the use of broad-spectrum insecticides that would otherwise have direct and indirect effects on natural communities dwelling near agricultural fields.The insecticidal proteins expressed in Bt crops are both highly specific and contained within plant, minimizing the possibility of any off-site effects due to spray drift. *Third*, herbicide tolerant crops facilitate a reduction in tillage, thereby reducing soil erosion, eutrophication and contamination of aquatic communities. The greatest threat to natural biodiversity comes from habitat loss, much of which is driven by agricultural demand. Increasing the productivity of land currently in production is necessary to slow this process. Other agricultural practices also can negatively impact natural communities through various off-site effects, including movement of fertilizers into aquatic systems and pesticidal drift. Reducing tillage and decreasing the use of pesticides can mitigate some of these impacts. GM crops can be a partial solution to several of these problems; GM crops enhance productivity, minimize off-site effects, and (in the case of herbicide tolerant crops) facilitate reductions in tillage.

APPLICATIONS OF BIOTECHNOLOGY TO VEGETATIVELY PROPAGATED CROPS IN AFRICA

In vitro propagation of vegetative crops is a second-generation biotechnology activity that can be successfully applied in developing countries. Cassava, yam, bananas, and plantains are important crops in Africa, and biotechnology has made significant contributions to their improvement. At IITA, biotechnology has been used for the last ten years as a tool for improvement of these crops. An extensive *in vitro* genebank of cassava and yam is maintained here for distribution worldwide. Tissue culture methods and cryopreservation techniques have been developed for these crops. In the case of yam, a molecular genetic linkage map has been developed, and putative transgenic shoots obtained. A polymerase chain reaction (PCR) test for yam virus II has been developed in a collaborative project with the Natural Resources Institute (NRI). This sensitive test has been useful in the programme to distribute improved varieties to farmers.

In the case of banana, RAPD markers linked to A and B genome sequences have been identified and tissue culture used for mass propagation. Several diagnostic tests have also been developed for detecting banana streak virus, in collaboration with the John Innes Centre and the World Bank. Research on regeneration and transformation techniques in cassava are being carried out at ARCRoodeplaat in South Africa and IITA in Nigeria. Here, techniques of organogenic regeneration, developed in Switzerland, have been successfully applied in both laboratories through a technology transfer activity.

TRAINING OF AFRICAN SCIENTISTS

Training of African scientists has taken place at two main training centers in Africa. One is at IITA, where training is a major activity of the Biotechnology Unit. Between 1990 and 1995 about 50 African scientists were trained and six students completed Ph.D. degrees.

The training activities included three annual workshops on the use of monoclonal antibodies for the detection of crop viruses, which were attended by 35 scientists from 19 countries.

The Life Science Programme of UNESCO has been very active in developing countries through their Microbial Resources Centers (MIRCENs) and Biotechnology Education and Training Centers (BETCENs). These programmes are providing training to scientists in Africa, enabling biotechnology to be applied to solve local problems. In 1995, the Biotechnology Action Council (BAC) of UNESCO established a BETCEN, at the ARC-Roodeplaat in Pretoria. Since 1995, a total of 180 scientists from 23 countries have been trained in basic and advanced tissue culture techniques, and applications of molecular markers.In addition to these short-term courses, 11 fellowships were provided for training periods of 2 to 3 months. The International Atomic Energy Agency

(IAEA) and the African Regional Co-operative Agreement (AFRA) have also been involved in supporting regional training initiatives in Africa. One of the more serious problems with training of African scientists is the lack of opportunities for African graduates once they return to their home-countries, and particularly after obtaining degrees in developed countries.Training gained in developed countries using hi-tech equipment to study esoteric topics does often not equip African scientists to return and contribute to growth in their own countries.

Some countries in Africa are now sending students for training at South African universities in the hope that graduates will be more willing to return to their home-countries once their studies have been completed.

FIELD TRIALS OF TRANSGENIC CROPS AND COMMERCIAL RELEASES OF GMO'S

In the period between 1990 and 1995 there were 25 field trials of transgenic crops in Africa. These involved a variety of crops and introduced traits. Of the field trials, 22 were performed in South Africa, 2 in Egypt and 1 in Zimbabwe. Africa has performed relatively few field trials of transgenic crops compared to the numbers of trials in the rest of the world for the time period 1986-1995. These are, North America (2,438 trials), Western Europe (796 trials), Asia-industrialized (86 trials), Asia-developing (62 trials), Latin America (204 trials), and Eastern Europe and Russia (36 trials).

Since this time, however, several other countries, for example, Kenya, Uganda, Namibia, and Cameroon, have introduced National Biosafety laws and regulations, and discussions are also being conducted in Mauritius, Zambia, Tanzania, Ethiopia, Nigeria, Ghana and C.te d'Ivoire. As a result, the number of field trials has also increased, and in South Africa between 1996 and 1999, there were 90 applications for transgenic field trials (M. Koch, personal communication). Field tests of transgenic crops have taken place since 1989 in South Africa under strict compliance of the South African Committee on Genetic Experimentation (SAGENE) guidelines. The GMO bill (Act 15 of 1997) has since replaced SAGENE.In 1998 in South Africa, there were commercial releases of two insect resistant yellow maize varieties. About 1000 hectares of transgenic maize resistant to the local stem borers, Busseola fusca and Chilo partellus, were planted in the 1998/1999 season. Another release of a genetically modified crop in South Africa has been in cotton. A recent trial release of insect resistant Bt-cotton showed a significant decrease in insecticide use and increases in yield between 17 and 24% for commercial farmers, and 28% for a group of resource-poor farmers. One woman farmer from a rural area in KwaZulu Natal said that she had made $5,000 more profit than she expected.

GENE DISCOVERY FOR DEVELOPING COUNTRIES

Over 50 years ago, Nicolai Vavilov alerted the scientific community to the value of conserving plant genetic resources to identify useful genes for breeding programmes. More recently, Tanksley and McCouch highlighted the usefulness of molecular tools to plant breeders.

In developing countries there is a need to build capacity in gene discovery, where in the absence of such programmes, many of the crop genetic engineering projects are "copycat" applications using existing genes from developed countries. These may not necessarily provide the most effective solutions to local problems in developing countries.

In addition to this, Africa contains an abundance of untapped, indigenous knowledge and genetic wealth, which could benefit developed and developing countries. There has also been concern expressed that the Biodiversity Convention does not clearly entrench adequate intellectual property right (IPR) protection for indigenous knowledge. There is also a school of thought that IPRs are inappropriate when applied to plant genetic resources in developing countries, since this benefits industrialized countries. A more practical approach, however, would be to build local capacity in developing countries, in collaboration with developed countries, in order to identify local genetic wealth, and then use this to address specific local problems.

The application of plant biotechnology techniques, in conjunction with conventional plant breeding and good crop management, can play a major role in ensuring food security and adequate nutrition in Africa. In this chapter we have provided examples of initial successes in the application of plant biotechnology techniques. Strong leadership, effective priority-setting, and adequate working opportunities for scientists are, however, required to provide incentives for the establishment of capable biotechnology groups.

Participatory extension approach programmes could be an ideal channel for the implementation of biotechnology products, as well as endowing resource-poor farmers with the confidence to develop and apply solutions to some of their problems. Africa's advantage is its valuable biodiversity, which can earn revenue for its people not only through tourism, but also through bioprospecting, if the value-addition remains in Africa. Programmes to build capacity in African laboratories to identify, classify, and utilize the continent's biochemical and genetic diversity will go a long way towards bringing the promise of biotechnology to Africa.

AN ENERGY TRANSITION FOR SARD AND FOOD SECURITY

During the last five years a new context for action has emerged in the international community, which calls for an energy transition in both developing and industrialized countries. For the OECD countries this transition takes form in the context of high consumption patterns that involve net resource transfers from South to the North, consume excessive resources and/or generate pollution.

In the African countries, a transition to sustainable energy systems is needed to accelerate the growth of basic food production, harvesting and processing. However, breaking the current energy bottleneck must also be sustainable (*viz.* environmentally sound, socially acceptable and economically viable). Such a transition involves a commitment to long-term developmental goals and requires innovative policy and technological solutions.

For Africa, an energy transition would be characterized by a move from the present levels of subsistence energy usage based on human labour and fuelwood resources, to a situation where household, services and farming activities use a range of sustainable and diversified energy sources. Obvious benefits are greater resilience in the production system, higher productivity, improved efficiency and higher incomes to farmers.

Environmental degradation, driven primarily by poverty, would be minimized. The investment required to make such a transition would not be significantly different from that required for conventional approaches.

However, the process of identifying needs and promoting investment in a range of technological options would be considerably different. The new energy situation would offer opportunities to reinvigorate the situation in many African countries which continue to cope with insufficient rural energy supplies.

Among the problems are the following:

- Price policies rarely reflect the energy needs of rural populations;
- Energy plans and agricultural programmes are not linked;
- Energy requirements for agro-activities are seldom quantified;
- Energy policies and plans in most countries do not focus on the agricultural and rural sectors, except occasionally, on an aggregate basis;

Unfortunately, the energy transition in rural areas will not occur under a "business-as-usual" situation. A concerted effort is needed on the part of the many actors influencing energy supply and demand patterns. One challenge is to reduce the barriers facing rural energy development which arise from a lack of policy and programme coordination between the rural and agricultural sectors, and the energy sector institutions.

The rural sector continues to remain outside the energy assessment and planning efforts which are normal practice for industry, commerce, transport. This is due, in part, to the small impact rural energy has on the national energy balance. Because of their meagre energy consumption and poor data, especially in the African context, the important role agriculture plays as a source of food and fibre, for foreign exchange and the relatively large percentage of GDP that is derived from the agricultural and rural sector, is ignored.

The dispersed and often non-monetized nature of rural energy also contributes to its neglect in planning and investment. Energy authorities rarely have an institutional or operational presence in rural areas and only a few agriculture and rural development programmes deal explicitly with rural energy requirements. This is due, in part, to lack of technical capability. However, a change in mindset is also needed among policy makers to recognize the potential economic and social gains to be realised from increasing energy supply in rural areas. These gains will translate into improved use and management of land resources by allowing more efficient use of resources and less degrading land-use practices such as fuelwood use.

National agricultural and rural development authorities, normally without any mandate regarding energy matters, are often incapable of negotiating their energy requirements with electricity utility companies and energy authorities.

Thus, a "vacuum" of responsibility and lack of guidance for energy interventions in rural areas seems to exist in most countries. No institution is actually "in charge" of energy for development of the rural and agricultural sector. This leads to low allocation of resources and investment for rural development and agricultural activities vis-à-vis other sectors of the economy. Since no single institution, either governmental, local or private could alone cope with all issues involved, a political interest, coupled with effective inter-institutional cooperation and collaboration is required.

Promoting food security by raising agricultural productivity and sustainable production systems will inevitably involve increases in energy inputs for water supply and management, plant nutrients, and agro-processing, to provide community lighting and drinking water, for cultivation. Consistent with the SARD framework mentioned earlier in this report is the need to shift the emphasis from single issue solutions to more integrated, sustainable approaches to development.

For example, pesticides alone are not sufficient or economically cost-effective in controlling most pest problems. Strategies now exist for many crops which involve understanding the pest life cycle, economic damage thresholds and the effects of cultivation practices which can greatly reduce or even eliminate the need for regular pesticide applications. IPM is an effective way of reducing production costs and avoiding the associated risk of pollution and contamination.Similar evidence of the benefits from integrated approaches exists for mineral fertilizers. Integrated plant nutrition strategies that use organic materials, leguminous crop rotation, and cultivation practices to maintain the optimal balance of soil structure and plant nutrients for agriculture are more beneficial economically to farmers than solely relying on mineral fertilizers.

The availability of adequate water resources for agriculture is essential for increased production. From 1960-1980, water resources in Africa have declined from 16.5 million cubic metres per caput to 9.4. By the year 2000 this is expected to 5.1.

It is projected that by the end of the 1990s six out of seven east African countries and all five north African countries bordering the Mediterranean sea (Algeria, Egypt, Libya, Morocco and Tunisia) will face acute water shortages; with six of the countries having less than 1 million cubic metres per person.

However, efficient use of this resource in Africa does not imply large scale, energy-intensive irrigation schemes. Small pumps have had an important beneficial effect on irrigation in some African countries of vegetable and even rice production. Where surface water is available this technology represents a well distributed and energy efficient option.

Previous experience in Africa has shown that the way in which the water resource is made available, both its price and mode of delivery, will determine whether the resource is used sustainably. Thus irrigation schemes should follow the principles established by the International Action Programme on Water and Sustainable Agricultural Development which takes into account the planning, development and management of water resources in an integrated manner. Also important is the potential for biomass energy conversion technologies. Residues from wood and agro-industries, purposely grown biomass and municipal solid wastes may play a major role in many African countries. The economic and social assessment of these options is needed to avoid disrupting employment and resource use. Local and global environmental benefits of biomass energy conversion must also be considered.

As a result of UNCED, efforts are underway to develop national Agenda 21's. These provide valuable entry point for countries to fully integrate rural energy requirements and potentials into their energy and SARD strategies. They involve multidisciplinary and convergent actions at both national and local levels, which draw together energy, agricultural and environmental knowledge, experience and policies. National and district rural energy strategies are needed to provide a common framework and plan to direct investment and pull together the efforts of various ministries such as Agriculture, Energy, Forestry, Planning and Agrarian Reform. NGOs, local community groups, and the private sector have an important role to play in such initiatives.

There are signs in some African countries such as Ghana, Morocco, Tanzania, Tunisia and Zimbabwe that new institutional and energy planning approaches are gradually emerging to improve the availability of rural energy supply for rural development.

Awareness of the constraints facing national and local authorities when trying to solve energy problems in rural areas is being increasingly converted into actual action. Decentralization of the decision making process and of energy production, enhanced social participation, institutional linkages, and the entry of new technologies are only some of the elements which will directly and indirectly influence a mobilization of efforts towards achieving Food Security and SARD.

BIOTECHNOLOGY IN THE PRODUCTION OF FOOD INGREDIENTS

As described in the Introduction, flavouring agents, organic acids, food additives and amino acids are all metabolites of microorganisms during fermentation processes. Microbial fermentation processes are therefore commercially exploited for production of these food ingredients. Metabolic engineering, a new approach involving the targeted and purposeful manipulation of the metabolic pathways of an organism, is being widely researched to improve the quality and yields of these food ingredients. It typically involves alteration of cellular activities by the manipulation of the enzymatic, transport and regulatory functions of the cell using recombinant DNA and other genetic techniques. Understanding the metabolic pathways associated with these fermentation processes, and the ability to redirect metabolic pathways, can increase production of these metabolites and lead to production of novel metabolites and a diversified product base.

BIOTECHNOLOGY IN DIAGNOSTICS FOR FOOD TESTING

Many of the classical food microbiological methods used in the past were culture-based, with microorganisms grown on agar plates and detected through biochemical identification. These methods are often tedious, labour-intensive and slow. Genetic based diagnostic and identification systems can greatly enhance the specificity, sensitivity and speed of microbial testing. Molecular typing methodologies, commonly involving the polymerase chain reaction (PCR), ribotyping (a method to determine homologies and differences between bacteria at the species or sub-species (strain) level, using restriction fragment length polymorphism (RFLP) analysis of ribosomal ribonucleic acids (rRNA) genes) and pulsed-field gel electrophoresis (PFGE, a method of separating large DNA molecules that can be used for typing microbial strains), can be used to characterise and monitor the presence of spoilage flora (microbes causing food to become unfit for eating), normal flora and microflora in foods. Random amplified polymorphic DNA (RAPD) or amplified fragment length polymorphism (AFLP) molecular marker systems can also be used for the comparison of genetic differences between species, subspecies and strains, depending on the reaction conditions used. The use of combinations of these technologies and other genetic tests allows the characterisation and identification of organisms at the genus, species, sub-species and even strain levels, thereby making it possible to pinpoint sources of food contamination, to trace microorganisms throughout the food chain or to identify the causal agents of foodborne illnesses. Monoclonal and polyclonal antibodies can also be used for diagnostics, *e.g.* in enzyme-linked immunosorbent assay (ELISA) kits. Microarrays are biosensors which consist of large numbers of parallel hybrid receptors (DNA, proteins, oligonucleotides). Microarrays are also referred to as biochip, DNA chip, DNA microarray or gene arrays and offer unprecedented opportunities and approaches to diagnostic and detection methods. They can be used for the detection of pathogens, pesticides and toxins and offer considerable potential for facilitating process control, the control of fermentation processes and monitoring the quality and safety of raw materials.

This conference deals with the application of biotechnology to food processing in developing countries. Biotechnological research as applied to bioprocessing in the majority of developing countries, targets development and improvement of traditional fermentation processes. We consider some areas specifically relevant to developing countries and list some key issues that should be considered by participants in this conference.

SOCIO-ECONOMIC AND CULTURAL FACTORS

Traditional fermentation processes employed in most developing countries are low input, appropriate food processing technologies with minimal investment requirements. They make use of locally produced raw materials and are an integral part of village life. These processes are, however, often uncontrolled, unhygienic and inefficient and generally result in products of variable quality and short shelf lives. Fermented foods, nevertheless, find wide consumer acceptance in developing countries and contribute substantially to food security and nutrition.How will applications of biotechnology to fermented foods impact on these socio-economic and cultural factors?

INFRASTRUCTURAL AND LOGISTICAL FACTORS

Physical infrastructural requirements for the manufacture, distribution and storage (*e.g.* by refrigeration) of microbial cultures or enzymes on a continuous basis is generally available in urban areas of many developing countries. However, this is not the case in most rural areas of developing countries.

Should research be oriented to ensure that individuals at all levels can benefit from applications of biotechnology in food fermentation processes, *i.e.* should logistical arrangements for starter culture development be integrated into biotechnological research targeting improvement of traditional fermentations? What is required for the level of fermentation technologies and process controls to be upgraded in order to increase efficiency, yields and the quality and safety of fermented foods in developing countries?

NUTRITION AND FOOD SAFETY

Fermentation processes enhance the nutritional value of foods through the biosynthesis of vitamins, essential amino acids and proteins, through improving protein and fibre digestibility; enhancing micronutrient bioavailability and degrading antinutritional factors. Many bacteria in fermented foods also exhibit functional properties (probiotics).

The safety of fermented food products is enhanced through reduction of toxic compounds, such as mycotoxins and cyanogenic glucosides, and production of antimicrobial factors, such as bacteriocins, carbon dioxide, hydrogen peroxide and ethanol, which facilitate inhibition or elimination of food-borne pathogens.

Are the nutritional characteristics (and safety aspects) of fermented foods adequately documented and appreciated in developing countries? Is there a need for consumer education about the benefits of fermented foods?

INTELLECTUAL PROPERTY RIGHTS (IPRS)

The processes used in the more advanced areas of agricultural biotechnology tend to be covered by IPRs and these rights tend to be owned by parties in developed countries. This applies also to biotechnology processes used in food processing. On the other hand, many of the traditional fermentation processes applied in developing countries are based on traditional knowledge.

In addition to biotechnology processes, microbial strains may also be the object of IPRs. For example, an era of massive private investment in biotechnology was initiated when the United States Supreme Court ruled in 1980 (in the Diamond versus Chakrabarty case) that a live GM bacterium (of the genus *Pseudomonas*, modified to degrade components of crude oil) could be patented. Many of the microorganisms associated with traditional fermentation processes in developing countries are unique. Issues of ownership will become increasingly important as bacterial strains are characterised and starter cultures are developed in developing countries.

How should food scientists, researchers, industry and governments in developing countries approach these issues?- A considerable volume of research into the development and improvement of fermentation processes is currently taking place worldwide. Are the research results from developing countries adequately documented? Who owns this information? Are cell banks being developed to protect microbial strains characterised in developing countries?

COMMERCIAL OPPORTUNITIES

Biotechnological innovations have greatly assisted in industrialising production of certain indigenous fermented foods. Indonesian tempe and Oriental soy sauce are well known examples of indigenous fermented foods that have been industrialised and marketed globally. The results of biotechnology research will lead to fermented foods of improved quality, safety and consistency.

5

Conventional Methods of Crop Improvement

INTRODUCTION

Pluses play a significant role in Indian agriculture because they provide protein rich diet to poor people of the country. They contain 20-29% protein *i.e.* about 2.5 times more than cereals. Inspite of release of large number of high yielding varieties, extension of irrigated areas and use of agrochemical, growth recorded in total production of pulses from 11.82 million tonnes in 1970-71 to 13.19 million tonnes in 1995-96 is considered to be non revolutionary. This resulted in decline in per capita availability of pulses to 37g/person/day. There are several reasons for stagnation in pulses production such as narrow genetic base, lack of genetic variability in useable genepool, inability to transfer increased input to increase output (seed yield) susceptibility to large number of biotic and biotic stresses, photoperiod sensitive and non synchronous flowering and maturity. It has been estimated that on account of insect pest complex infesting pulse crops alone, nearly 2.0 to 2.5 million tonnes of pulses is lost annually. The monitory value of this loss will be approximately Rs.3000 to 3750 crores.

Conventional methods of crop improvement have not improve productivity of pulses with significant pace due to several reasons. These methods are slow, expensive and time consuming. Besides lack of reliable resistant sources, undersirable linkages and long gestation period are considered to be the main limitation of these methods. Hence, there is dire need to introduce innovative biotechnological approaches to enhance pulse productivity to meet out the market demand.

Biotechnology and genetic engineering has emerged as potential tool in recent year for genetic manipulation of crop plants. The said technology promises to improve crop productivity through complementing traditional breeding by decreasing dependence on harmful chemicals pesticides. Fertilizers and antibiotics etc. The tools of biotechnology are now available to plant breeders and can be used effectively to supplement conventional practices in improvement of pulse crops. It is expected that products of biotechnology will be ecologically sage, economically feasible and technologically expectable to Indian farmers. Effort have been successful to develop several intergeneric and interspecific hybrids using embryo/ovule/ovary/endosperm culture in grain legumes, Several useful genes imparting resistance to various biotic and biotic stresses have been successfully introgressed on wild relatives of lathyrus, lentil, chickpea, pigeonpea urd bean and mung bean. The resistance to Aschochyta blight to chickpea has been successfully transformed from wild species Cicer judaicum to elite chickpea genotype PDG 84-10 at IIPR, Kanpur.

The techniques of in vitro mutagenesis in vitro selection and somaclonal variation are being also successfully employed to various pulse crops to develop promising genotypes resistant to various biotic and a biotic stresses. This technique has resulted success in developing Aschochyta blight resistance in chickpea. Further early results in vitro selection for salinity tolerance in chickpea is also very encouraging at IIPR, Kanpur. Similarly, the technology of cell suspension culture followed by selection of mutants has also been used successfully in major pulse crops at different centres in the country.

Identification and characterization of pathogens and pests is problematic and always pose challenge to plant pathologist and breeders in varietal development of pulse crops. The techniques of molecular diagnosis and DNA finger printing has facilitated resulted easy diagnosis and the precise characterization of pathogeny isolates and insect biotypes at molecular level. This has increased the efficiency of breeding programmes. Important pathogent like Aschochyta rabi and Fusarium oxisporum of chickpea have already been characterized using DNA fingerprinting techniques.

In a new world order and changed senario the characterization of germplasm and plant material is very essential to protect out natural wealth. The techniques of DNA fingreprinting have been proved very handy in protecting the intellectual property rights. These techniques are now being successfully used in pulse crops to characterize and catalogue the accessions of gene bank to protect the genetic wealth.

DNA marker techniques are the new innovations and have tremendous potential in increasing the efficiency of conventional breeding programmes. DNA marker technique can be used for tagging important genes which can subsequently be used for early, easy and precise selection in segregating of promising plants populations. DNA markers have been found linked to wilt resistance in chickpea, powdery mildew resistance in peas and nemarode resistance in mung bean and urd bean. Transgenic technology which involves insertion of foreign DNA sequence has tremendous potential for improvement of crop plants. Efforts are on to develop transgenic plants resistance to pod borer in chickpea and pigeroxpea using Bt. Crystal protein gene from a soil bacterium. Early results obtained at various centres in the country are very encouraging. Similarly, efforts are also being made to develop transgenic urdbean and mung bean using coat protein gene. Current status and future strategies for genetic manipulation of pulse crops using biotechnological tools.

AREA, PRODUCTION AND PRODUCTIVITY OF MAJOR PULSES

India is producing 14.76 million tons of pulses from an area of 23.63 million hectare, which is one of the largest pulses producing countries in the world. However, about 2-3 million tons of pulses are imported annually to meet the domestic consumption requirement. Thus, there is need to increase production and productivity of pulses in the country by more intensive interventions.

ONGOING SCHEMES ON PULSES DEVELOPMENT

Presently, pulses development programmes are being implemented through the Centrally Sponsored Schemes of NFSM-Pulses, Integrated Scheme on Oilseeds, Pulses, Oil Palm and Maize (ISOPOM) and Integrated Crop Development Programme (ICDP)-Pulses under Macro Management Mode. While ISOPOM is implemented in all the 433 districts of 14 States, NFSM Pulses is being implemented in mission mode in 171 districts of these states. In addition, 15 districts of Jharkhand and 10 districts of Assam have recently been included under NFSM-Pulses for implementation with effect from 1st April, 2010. ICDP-Pulses is implemented in states other than the 14 NFSM/ISOPOM States. NFSM-Pulses cover about 80% of the total area under pulses in the country.

BIOTECHNOLOGICAL APPLICATIONS FOR THE IMPROVEMENT OF INDIAN TOBACCO

India is one of the top 10 countries cultivation tobacco for the part ten decades both for domestic and international market, India is the third largest producer of tobacco in the world with an output of approximately 520 m kg. Of which FCV tobacco account for 120 m kg. The economic importance of to Indian exchequer is peogressively increasing in the last 10 years. Tobacco generates 10% of total excise revenue and tobacco 4% of the value of total agricultural exports, India has the unique distinction of growing different types of tobacco in different Agroclimatic zones. Tobacco provides direct and indirect livelyhood to 26 million in India including 6 million farmers and workers. Many small and marginal farmers were benefited by growing tobacco. Research efforts to identify alternative crops to tobacco indicated that no other crops is as remunerative as tobacco. Stability of the worldwide demand of the crop, the crops hardiness and ability to grow in climate and soils non-suitable for other crops, relative ease in transporting tobacco and tobacco's profitability compated to other crops are incentives for farmers growing tobacco thus making tobacco one of the most appealing cash crops to farmers. Tobacco crops is considered as major employer worldwide. Despite technological progress tobacco cultivation remained as one of the most labour intensive crops among all other agricultural crops. Women in particular and skilled labourers, who would otherwise have little chance of employment are able to earn living by working in tobacco cultivation. WHO in its report has acknowledged that in developing countries tobacco can create vast employment. In view of its great role in the economy of the country and farmers, further improvement of tobacco crop through Biotechnological tools will benefits them a great deal.

Biotechnological applications: Scientific system of tobacco cultivation in India was introduced with the establishment of Central Tobacco Research Institute (CTRI) at Rajahmundry in 1947. Tobacco research in India over the past 50 years through CTRI has made significant contributions to the economic upliftment of tobacco farmers by identifying high yielding varieties and suitable package of practices This is reflected in the good reputation for Indian Tobacco in the International market, as leaf of desired quality traits as per the specific requirement of traders is produced. Initially, conventional breeding methods were utilised for improving the yield and quality of tobacco. However, later application of biotechnology tools for supplementing the conventional tobacco improvement programmes or initiated.

Different biotech techniques like anther culture, fertilized ovule culture micropropagation, protoplast culture somatic hybridization and also genetic engineering are being utilised. These techniques are helpgul is strengthening the conventional breeding programmes where ever they are week and failing to serve the purpose.

YIELD IMPROVEMENT

Yield improvement through conventional breeding programmes takes fairly long time. This can be reduced substantially (3-4 years) by culturing anthers (Anther culture). Of elite lines from early breeding populations in tissue culture. Through anther culture several dihaploids are produced and evaluated every year. An improved line D1 was developed, in this way, which is superior in yield. Efforts to incorporate budworm resistance through someclonal variation in this line is in progress. Stable resistant D1 someclones are there in field testing stage.

Resostance to Biotic and Abotic Stresses

Wild species of tobacco possesses resistance to various tobacco pests and disease. It is difficult and in some cases impossible to transfer this resistance from wild species to cultivated tobacco through conventional methods. In such cases, inter-specific hybrids were rescued through fertilized ovule culture and mass multiplied for further testing by micropropagation, In this way, interspecific hybrids possessing resistance to leaf spot diseases, black shank, root knot nematodes leaf eating caterpillar, aphids, budworm etc. were developed and evaluated for their resistance. Thus, stable pest and disease resistant lines were identified through Biotechnological tools. Efforts to transfer desirable characters from wild species through somatic hybridization is in progress.

Somaclonal variation was attempted to fortigy tobacco lines resistance to white fly (Bemisia tabci Genn.) (leaf curl) and budworm (Heliothis armigera Hb) and the results are encouraging. Somaclonal variation is highly effective to create variation when desired variation is impossible to get through conventional breeding methods and also variation is required only for one character. Transgenic Bt tobacco cultivars processing resistance to leaf eating caterpillar (Spodoptera litura F) and budworm (Heliothis armigera Hb) were developed. Production of transgenics will give scope for transferring desirable characters from different species and genera.

Preliminary attempts were made to screen lines for drought tolerance in tissue culture. Such efforts will save expenditure on extensive field evaluation and work can be carried out in the off season also. Transgenic tobacco having wheat desiccation tolerant protein were evaluated water stress tolerance.

Herbicide Resistance

Orabache cernua os a root parasite of tobacco causing heavy loss in yield and quality. Glyphosate at lower does was found to kill Orabanche, but it is phytotoxic to tobacco. Hence research on production of herbicide (glyphosate) resistant tobacco lines through tissue culture was initiated. When farmer grows herbicide resistant lines in the field he can use berbicides without any harmful effect on tobacco, for the effective control of weeds and Orabanche.

Conservation of germplasm: Germplasm is the valuable source for various characters that are useful in future breeding programmes. This can also be conserved under in vitro conditions for long time. Also lines that cannot be maintained under field conditions due to non-flowering and sterility can be effectively maintained under tissue culture conditions. Interspecific F1s, mammoth mutants, in vitro rescued hybrids protoclones, BT transgenics various mutants are routinely being maintained.

Future Thrusts

Biotechnology can help Indian tobacco farmer in many ways, Particularly in tobacco improvement research and seed purity disputes, if offers a large scope. Research on tobacco improvement benefit farmer as he can get high yielding and pest and disease resistant tobacco for cultivation. In research various aspects like in vitro conservation and molecular characterisation of germplans, isolation, characterisation and transfer of desirable genes from across genera, molecular aided selection in breeding programmes, someclonal variation can play a great role. Utilising Biotechnological techques tobacco lines that can yield higher amounts of phytochemicals which are useful in the production various drugs and pesticides can also be developed. In recent times many disputes about seed purity among farmers are coming up. At present morphological criteria is only being used for varietal identification. As morphology is highly subjective in nature and also varies with the type of environment the plant is grown, it is becoming highly difficult to solve the disputes. Under this situation molecular characterisation of cultivated varieties will help to identify correct varieties beyond doubt.

In future, giving emphasis on the above said lines will go a long way in benefitting the India tobacco crop as well as the farmer. However, as biotechnology got wider applications, any thing that put the farmer and environment at disadvantage need to be discounrages.

BIOTECHNOLOGICAL APPROACHES TO WEED MANAGEMENT

Weed management is essential for good quality and quantity of food production. The presence of weeds in general may reduce crop yield by about 30% or more. Being potential pests, weeds have to be removed employing mechanical, chemical cultural and biological means. Weed management through application of chemicals (herbiciedes) is, through effective, not eco-friendly. The herbicides may pollute the environment and many alter the natural equillibrium. They also endanger and alter plant and microbial biodiversity. Effective non-chemical weed management may include various biotechnological approaches using plants and/or their products. These are useful both for the natural equilibrium and the environment. In addition they also condition the soil and improve crop production. Some of the recently evolved biocontrol and biotechnological methods of weed management developed and tried at the National Research Centre for Weed Science, Indian Council of Agricultural Research, Jabalpur are outlined in the following text for ready use for trials and for working our their cost effectiveness.

Management of Phalaris Minor in Wheat by using Fungi

At present Phalaris minor is one of the most predominant weeds in wheat fields. The Weed spreads through seeds. The weed has developed resistance against popular herbicides like isoproturon. Biotechnological approach using fungi offers an alternative possibility of management of the weed through biocontrol.

(i) Control of Phararis minor using Trichoderma viride: Trichoderma viride is used as a seed dressing fungi in various crops including in wheat for management of seed borne/soil borne pathogens. Interestingly, the fungus has been found to inhibit germination of seeds of Phalaris minor. Application of Trichoderma, viride grown in saw dust and neem oil cake as a base consistently inhibits germination of phalaris minor seeds and reduces vigour of the seedings of the weed. At the same time it improves yield of wheat by about 5%. The fungus has potential of controlling phalaris minor. Soil application of the fungus, wheat seed treatment and spray of the fungus were all effective in controlling the weed.

(ii) Control of Phalaris minor by gilocladium virens: Control of Phalaris minor by ?Gliocladium virens has also been found to have potential for control of Phalaris monor by inhibiting seed germination in a similar manner as does the Trichoderma viride.

MANAGEMENT OF PARTHENIUM (PARTHENIUM HYSTEROPHORUSL) BY MARIGOLD (TAGETES PATULAL)

Parthenium is an obnoxious weed of worldwide occurrence, harmful to human and animal health, agriculture, environment and the natural biodiversity. The weed reduces yields of crops and grasses. When shown in parthenium infested field or area, marigold inhibits seed germination and growth of the weed.

It also reduces population of the weed. In the next generation, reduction of parthenium population is as height as 85-100%. This has been found primarily due to competition and through all elopathy (affecting parthenium plants by releasing chemical substances including thicophene into soil through roots.)

Built up of the phytotoxic substances of marigold origin prevent parthenium seed germination, growth of plants, and flower and seed production completely in 2-3 years. This technology appears to be economically rewarding for marisgold flowers have good market. This can be achieved simply by spraying marigold seeds over parthenium infested area. The technology is simple, effective, economically rewarding, and ensures self perpetuating control of parthenium in wastelands. Marigold to parthenium plant ratio of 0.5 to 4 appear effective for near complete control of parthenium. Adequate moisture availability in soil facilitates establishment of marigold in parthenium infected area.

Management of Parthenium by Fungi: Two fungal species namely Gliocladium virens and Trichoderma viride which have been widely used as seed dressing to present parthogenic damage have also been found effective for parthenium control.

(i) Control of parthenium by Gliocladium virus: Glioladium vieus and 10% neem oil separately as well as in combinations spray control pathenium and other broad leaf weeds like chenopondium album, Melilotus alba and Medicago sp. In wheat crop. The fungus could be explored as a self perp control measure for non-cropped area also.

(ii) Control of parthenium by Trichoderma viride and need oil: Thrichoderma viride and neem oil separately as well as in combinations spray control parthenium under field conditions. The fungi has potential of becoming a self perpetuating means for control of parthenium under non-cropped area. Trichoderma viride also acts as an antidote for parthogens of parthenium that can possibly be used as biocontrol agents. If any of the parthogens attack crops, the Trichoderma virid can be used to overcome the problem.

MANAGEMENT OF PARTHENIUM BY UTILISING AS A SOURCE OF NUTRIENTS

Parthenium is able to extract nutrients even from nutrient deficient soils in which it grows. It has very height levels of nitrogen (3%), phosphorus (0.2%), potassium (4.5.%) and other macro and micro-nutrients. Uprooted, dried and cut to small pieces or powdered parthenium plants at preflowering stage can be applied directly to crop fields.

Apart from increasing crop growth and yields, the parthenium residue conditions soil with the organic supplement. Keeping safety considerations, it may serve as an incentive for participation of public in its ecofriengly control. Weed management through mid-summer irrigation and covering the beds with black polythene: Weed problem in most cases is due to seeds of weeds present in soil seed bank which germinate before, with or after planting the frops. Mid summer irrigation and covering the beds or fields with black polythene thoughout the summer kill most of the readily geminable seeds in soil seed bank. Polythene covering can be removed after summer before planting the crops.

The treatment may provide near complete weed control. This technique is suitable for areas where summer temperature exceeds 400 c. Weed management through black polythene stripe mulching Where crop is grown in rows, inter row space could be covered with black polythene mulching. This kills weeds appearing in the area due to interception of light necessary for survival of the plants. The are many other ways of biotechnological management of weeds such as bioling water jet treatment of soil, by intensive cropping, by increasing plant density and chosing allelopathic crop species etc.

BIOTECHNOLOGYCAL APPLICATION

Techniques such as embryo transfer, dilution of semen and its cryo preservation and artificial insemination (A.I.) are well developed and expertise in these fields is available. This technology is extensively used in the case of cows and to some extent Buffaloes. The main lacuna in the case of sheep and goats is the non availability of improved breeds which could be utilised. As these animals are owned by the poor strata of society and there is a mistaken prejudice that they are responsible for environmental degradation no sustained efforts have been made to improve indigenous breeds and thus make available genetic stock for upgrading. In spite of this the numbers of these animals have been increasing rapidly (*2.5% per year in the case of goats) due to an increasing demand for meat. Our breeds have developed over the centuries to survive and adapts themselves to harsh environmental conditions but do not universally possess ability to respond to better management and nutrition.

Outstanding work in the case of the Sirohi goat was carried out from 1981 to 1993 by the Indo-Swiss Goat Development and Fodder Production Project in Rajasthan where they showed that outstanding animals are available within this breed. However, before the impact of this project could be felt the project was discontinued.

Goats

Out institutes is a very small NGO with only three technically qualified staff, one of who is in an honorary capacity and one on deputation from the Government of Maharashtra. We decided to initially concentrate on crossbreeding using the outstanding South African Boer Goat the worlds only meat goat as an improve breed. With assistance from the Government of Australia we imported 20 embryos of this breed from Australia implanted them in local goats and now have a small flock of Boer goats.

This goat unlike many temperate breeds of Dairy goats have proved to be extremely hardy, producing excellent crossbreed when crossed with local goats. In order to utilise this germplasm effectively we have developed and modified the techniques of semen cryopreservation so that it can be done with low cost equipment producing pellets instead of straws. We have also improvised. A.I,. techniques using cervical insemination to get better conception rates. We have so far trained over 50 trainees among them farmers. We have supplied pure Boer goat frozen semen to Governmental agencies in Orissa, Tamil Nadu as well as to may private goat rearers in Maharashtra.

Our veterinarian is capable of carrying out multiple ovulation and embyo transfer if there is need for rapid expansion of our flock. However, this technology is expensive and only to be used if costs justify its use.

A I technology in goats has been spread to near by villages where fresh diluted semen as well as frozen semen have been used to produce 800 crossbreeds over the past five years. This technology is viable and easily transferable but unfortunately dependent on the sincerity of the inseminator.

Sheep

We were unable to identify a suitable improve breed to be imported which had the most desirable but greatly neglected attribute of prolificacy. We did locate a prolific and unusual breed in the hot humid sundarbands of West Bangal called the Garole. Up to 1993 this breed had remained unnoticed by the community of sheep breeders. We did however import a small flock of Awassi Dairy Sheep from Israel. This breed has produced an outstanding crossbred animal when crossed with the Malpura of Rajasthan. This crossbred has been developed by the Central Sheep and Wool Research Institute (CSWRI) using Awassi rams obtained from us. The Australian Centre for International Agricultural Research has sponsored a collaborative project between the University of New England, Australia, The national Chemical Laboratory, Pune and the Nimbkar Agricultural Research Institute (A SISTER NGO of Maharashtra Goat and sheep Research & Development Institute), Phaltan entitled," Prolific Worm Resistant Meat Sheep for Maharashtra, India. This project is making full use of Biotechnology. We are comparing three breeds of Indian sheep the Grole, Deccani and the bannur. We hope to evaluate lamb production including prolificacy and resistance to parasites and try and determine the genetic basis of these differences by DNA analysis. We also hope to account for the interaction between measurement of resistance to parasitic nematodes and environmental effects. The project has completed one year and the results so far are encouraging.

Our ultimate objective is to provide shepherds with rams incorporating the prolificacy of the Garole, the growth potential of the Awassi and the meat quality and hardiness of the Bannur. A very important but at present neglected application of Biotechnology which requires considerable investment is in the field of disease diagnosis and vaccine preparation. Once productive animals that respond to inputs are available and the rearers can afford to pay for the above services I am confident that the private sector will enter this field to provide this input. It is imperative to review the cost benefit ratios of Government institutions and their contribution to the Goat and Sheep industries I think you will find that funds given to NGO's are more productive. There is much talk of helping NGO's but real assistance only seems to come from foreign aid agencies which is unfortunate.

TECHNOLOGY OF CROPPING SYSTEM
CROPPING SYSTEM

A cropping system may be defined as a community of plants which is managed by a farm unit to achieve various human goals. The latter include food, fibre and other raw materials, wealth and satisfaction. Conway encompassed these multiple goals in the phrase 'increased social value'. Farmers are part of the system. They are able to set, or modify, their own goals, so two farms with identical climates and soils may be managed with different aims to achieve a different mix of outputs. IRRI gives a more detailed definition of a cropping system: '.the crop production activity of a farm. It comprises all cropping patterns grown on the farm and their interaction with farm resources, other household enterprises and the physical, biological, technological and sociological factors or environments". The cropping system is one of three sub-systems likely to be present within a broader farming system.

Food security is the most basic output from a cropping system. For a subsistence household, all other outputs are secondary. For corporate agriculture, under a sophisticated cash economy, food security is established and other goals, such as profit, predominate. Thus, the first property of a cropping system is its effectiveness; that is its ability to meet the needs of the goal-setter (the farmer or corporate manager, etc.). The effectiveness of a system is a subjective attribute which should be assessed by the goal-setters and their society. It can be estimated or quantified through surveys using ranking techniques or by econometrical devices such as shadow pricing and opportunity costing.

To illustrate the importance of assessing the effectiveness of a cropping system. Here, in a survey, 270 Indonesian households were classed into six types (*e.g.,* aiming to be self-sufficient in food, wage-earning or in commercial agriculture) and, depending on their diet and the possibility of their diet being inadequate at some times of year, they were assigned to four classes of 'food insecurity'. Households managing cropping systems for self-sufficiency occur in all categories of food insecurity.

A successful household whose cropping system supports them in a 'food secure' state, will have different opinions on the effectiveness of their cropping system from a neighbouring 'food insecure' household that also aims at self sufficiency. Furthermore, it is likely that their attitudes to risk-taking and change are likely to be quite different. This is in part because of the difference in effectiveness of the two systems. One family may be open to innovation and the other perhaps caught in a spiral of conservative thinking and poverty.

Production is a most obvious output and measure of the activity of a cropping system. It can be measured as the biological or economic output from the system, for example as the grain or cash generated. It is implicitly an output from the activity of one or more management units (*e.g.,* families). It is also a measure of the efficiency of the management of the cropping system and can be related to productivity-measured as output per unit of input (land, labour, capital, energy). Conway (1985) defines other ecosystem properties as sustainability, stability and equitability. Of these, sustainability, the subject of this *Bulletin,* is the most difficult to define and estimate.

FAO (1989b) states that the goal of sustainable agriculture is to 'maintain production at levels necessary to meet the increasing aspirations of an expanding world population without degrading the environment', and that sustainability 'implies concern for the generation of income, the promotion of appropriate policies, and the conservation of natural resources'.Similarly, sustainable agriculture and rural development are defined as 'the management and conservation of the natural resource base, and the orientation of technological and institutional change in such a manner as to ensure the attainment and continued satisfaction of human needs for present and future generations'. Conway defines sustainability as the ability of an ecosystem to maintain productivity when subjected to a major disturbing force. He is thus forced to draw a continuum between stability (in the face of minor, expected forces) and sustainability (in response to major forces). He also indicates that to assess sustainability, one needs a measure 'of the effectiveness of internal adjustments that agro-ecosystems make in response to stresses and shocks'. A more pragmatic definition of sustainability, with which most agronomists would agree, is that 'a sustainable cropping system... is one which maintains resources, such as soil and water, while providing an adequate and economic level of production, both now and for generations to come'.

Similarly, a sustainable system is one in which:

- Resources are kept in balance with their use through conservation, recycling or renewal.
- Practices preserve agricultural resources and prevent environmental damage to the farm and off-site land, water and air.
- Production, profits and incentives retain their importance, because not only agriculture needs to be sustained, but so do farmers and society.

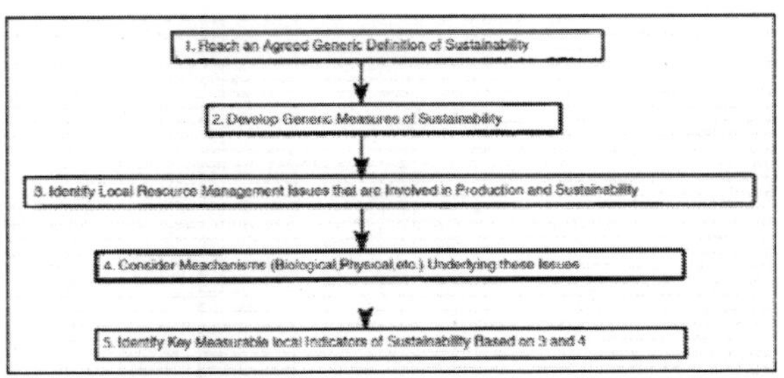

Fig. Measuring Sustainability

Conway's definition allows for levels, or degrees, of sustainability but constrains assessment of sustainability to measurements after shock. Hoare and others define a sustainable system in terms easy to relate to, but often do not accommodate degrees of sustainability, which is unhelpful for management and policy development.

Here, Conway's concept is accepted that it is useful to consider the possibility of various degrees of sustainability, and the widespread, but rarely developed, implication that sustainability should be measurable in either biological or economic terms (just as in measuring productivity). It follows that a biologically sustainable system may not be economically sustainable. This is witnessed by land in various parts of Europe and elsewhere that has gone in and out of cropping on several occasions during historic times depending on economic conditions.

Here, sustainability is defined consistent with FAO, as the ability of a cropping system to maintain productivity over a long term. Since, this definition implies also maintenance of resources such as soil, it requires that all participants (farmers, researchers, extension specialists and policy-makers) agree upon appropriate measures by which the sustainability of particular cropping systems are assessed.It is important that everyone involved in an assessment of the sustainability of a cropping system should agree on an inventory of resources which encapsulates its essential features. From this, the most critical attributes and, in some cases, those most easy to estimate, can be identified. The type of resource inventory that might be agreed necessary to define the sustainability of a semi-arid cropping system. Defined critical resources will differ for other systems. After defining the critical resources, it is necessary to identify those indices of sustainability that are easily and relatively-cheaply measured.

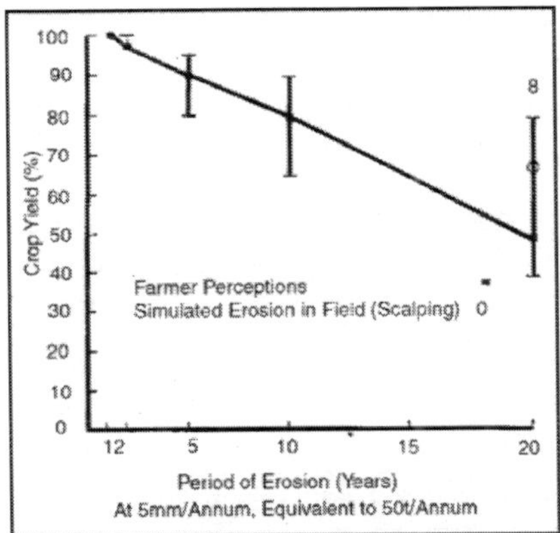

Fig. Crop Yield Declines with Continuing Soil Degradation and Erosion.

If a system is not sustainable, remedial action depends on agreement on the extent of non-sustainability and the changes in inputs to make it sustainable. Some research is encouraging here: for example, Rickson *et al.* find reasonable agreement between farmers' perceptions of the impact of continued soil loss on sustainability, as estimated by the extent of decline in wheat yields.Stability (or variability) is the constancy of productivity in the face of small disturbances such as seasonal or year-to-year fluctuations in weather.

It is a measure of the spatial- or year-to-year robustness of a farming practice and reflects not only variations in crop response to weather and soil type, but also variations in prices of inputs (such as labour and pesticides) and of the product. Last, but not least, equitability is the property that reflects how the resources and outcomes of the cropping systems are shared among the population.

EVOLUTION OF FIELD CROP ECOSYSTEMS

Various terms used in the sections that follow are defined in Appendix 1. Most of the definitions are taken from IRRI and ICRAF. Cropping systems have evolved over historical time along a catena of increasing energy inputs (including mechanisation), reduced biological diversity and increased risk or instability.

Boserup (1965) developed the thesis, forwarded and embellished by others, that population growth is a precondition for agricultural change. Subsistence communities in semi-arid regions depend on herding livestock (most common in semi-arid grasslands and woodlands) or shifting cultivation. In the latter, the cultivated plot is moved to new land when fertility is depleted or the plot becomes too weedy.

Boserup's thesis, backed by worldwide experience, is that increasing population, in places partly caused by immigration (*e.g.*, settlers moving into sub-Sahelian Africa), either:

- Reduces community mobility until the villages become static and maintain permanent fields;
- Already semi-permanent villages are so deprived of hunting/foraging grounds that they too need to develop permanent cropping systems. As the fallow period between crops is shortened and the fields become more permanent, two strategies are available;
- Extensification and/or increased mechanisation. For example, use of draught animals or tractors;
- Intensification including use of shorter fallows, perhaps coupled with use of unused poorer land. This is accompanied by increased complexity of cropping, *e.g.*, household gardens, and environmental modifica-tion, *e.g.*, irrigation.

Okigbo comprehensively described these changes in cropping systems, and their implications for sustainability.

CROPPING REQUIRES LIVESTOCK FOR DRAUGHT

Extensification of cropping requires livestock for draught, or mechanisation. Extensification may encourage supplementary pasture systems outside, as well as part of, the cropping systems themselves. It is also possible to develop extensive cropping without livestock. Agriculture in northeast Argentina and parts of Brazil, for example, has shifted directly from uncultivated grasing land or unused bush to extensive, mechanised cropping. Where cropping and livestock farming develop together, there are options as to their integration. In semi-arid Australia, mechanised extensification of 60 million ha of arable land has seen little (long term) change in the ratio of crop to pasture. Pasture land has increased in area since, about 1930 to stabilise at about 28 million ha in 1970.

Though livestock are not used for draught, this extensification has maintained a close link between crops and pasture in an integrated ley farming system. In contrast, in the semi-arid Near East, extensive cropping has been developed by private landholders without livestock. Here pasturage remains largely in the public domain with nomadic herds and common grasing land. Thus, though Australia and the Near East both have crop and livestock subsystems, in Australia they are integrated and managed by one family or corporation but in the Near East they may be separate and owned by distinct social groups.

INTENSIFICATION

Intensification involves biological or environmental modification. Biological modification occurs by increasing the complexity of a local cropping system by the development of multi-storied household gardens, relay cropping or mixed cropping. It is done by individual households. It increases labour inputs, diversifies crop management (though not necessarily crop diversity which may be greatest in hunter-gatherer systems), and usually increases the range of livestock (chickens, ducks, goats, fish and sometimes cattle are common in Javanese household gardens). It may reduce insect herbivores. Importantly, it creates food and income stability. Biological intensification can also be carried out on a larger, usually corporate scale. Multi-story plantation crops, such as beans under coffee, alley cropping and agroforestry, seek high outputs per unit area and stability in ways analogous to complex household gardens. Such intensification appreciably modifies the micro-environment as a secondary outcome. For example, tree legumes were planted along fence-lines in semi-arid Ethiopia in the late 1980s to provide quality forage for cattle and income from the sale of seeds. The trees provided shade and habitats for birds, which became more prized by villagers than the forage.Intensification can also be carried out by deliberate modification of the environment, most obviously through irrigation.

Intensification, whether by biological or environmental modification, generally gives:

- Greater diversity of land use.
- Some intensively-used cropping land, usually in areas with favourable hydrology, fertility or social attributes (*e.g.,* prior settlement, easy transport).
- Some under-used or unused land. Cropping systems in Mediterranean lands often show these features, though there under-used land sometimes reflects lack of population pressure.

It is notable that extensification of cropping systems can create relative uniformity but intensification creates diversity. Neither, however, has a predictable social or environmental outcome. Extensification, often using monocultures, can be inherently stable and sustainable, or not. The social distribution of its benefits may be egalitarian (giving either uniform prosperity, or uneconomic, socially-deprived family farms), or it may be hierarchical with economically-strong landowners and a landless class of labourers and farmhands. Intensification at the household level benefits the family directly; intensification through village cooperatives or corporations may be advantageous or create negative social and environmental outcomes.

GROWING SEASON OF CLIMATES AND DRYLANDS

FAO (1987) uses the term 'drylands' to describe climates with fewer than 120 days growing season. These are divided into 'arid drylands' with less than 75 days growing season, and 'semi-arid' areas which have from 75 to 119 days growing season. Seasonal changes in indices of crop growth in tropical drylands, classed by Hutchinson *et al, as* having climate types I (hot, seasonally wet/dry: uppermost row) or E (warm, seasonally wet/dry: middle row), and in temperate drylands (lowermost row). Growth index () is a dimensionless index from zero (no growth) to 1 (maximum growth).

Though most semi-arid lands border the arid or desert climates of the world, they support a diversity of cropping (mostly with annual crops). Because of this diversity their climate should be quantified and any variations within them delineated. Such drylands have either one or two wet seasons. Bimodal rainfall regimes are characteristically found in latitudes between the wet tropics and the unimodal more temperate, semi-arid zones. Their distribution, however, is governed more by pressure systems and land masses. For example, bimodal patterns are rarely found in east Asia. In the tropics, the wet season(s) coincide with a high solar angle except for a relatively small area in east Brazil. The tropical drylands are commonly called monsoon (in Asia) or savannah (in Africa) although there is confusion over these terms. Koppen uses 'monsoon' in a restricted sense, as a climate with a dry season but able to support rainforest. In temperate areas, there is usually a single wet season that coincides with a low solar angle (winter). These wet-and-dry climates are generally called Mediterranean and, on the drier margins, semi-arid.

The tropical drylands (to use FAO's term) are categorised I and E by Hutchinson *et al.,* (1992); and the temperate drylands as E, D and C on the basis of seasonably of available water and temperature. This delineation of climates is based on calculations of the relative constraints of soil water, temperature and solar radiation on growth of temperate and tropical crops. Each of the indices ranges from zero (no growth) to 1 (climate optimal for growth). The water, temperature and radiation indices are multiplied together to calculate a growth index, which is an estimate of the overall suitability of the location for crop growth in a particular week or month of the year. Hutchinson's climate classification, unlike more traditional, geographically-oriented classifications, reflects seasonal fluctuations of environmental factors related quantitatively to crop growth. Examples relevant to the tropical drylands and climatically-similar wet-and-dry temperate areas.

Tropical drylands that support cropping have mean annual water indices as low as 0.29, but have a growth index (the product of the water, temperature and radiation indices) during the highest quarter (three-month period) greater than 0.3 (commonly 0.3-0.8). This is sufficient to support good growth of a single crop. Zone I coincides with parts of the Aw, Bsh and Cwa types of Koppen (1900) and Trewartha and corresponds with the 1.3, 1.4, 1.5, 1.7 and 1.9 zones of Papadakis. Hutchinson's E zone, which is not as hot coincides with parts of Koppen's Cfa, Cwa and Bsh types and Papadakis' zones 1.7, 2, 4.2 and 5 in the tropics.

PREDOMINANT SOILS OF THE SEMI-ARID CROPLANDS

The predominant soils of the semi-arid croplands are Luvisols and Calcisols in Africa and Australia, Mollisols and Luvisols in temperate areas of the Americas, Ferralsols in wet-and-dry tropical America, Luvisols and Chernozems in temperate Eurasia and Calcisols and some Acrisols in wet-and-dry tropical Eurasia. The climatic distribution, area and chief limitations of the main major soil orders. All these soils occur in seasonally wet-and-dry climates, although the Acrisols are not extensive in drylands. Lixisols, most widespread in potential cropping areas within seasonally wet-and-dry climates, have poor water-holding capacity, hard-setting surface horizons with a risk of crusting, run-off and erosion.

The topsoil texture varies from loamy sand to loam and clay, but the hard-setting properties and relatively poor infiltration and drainage (compared with, say, Acrisols) mean that crop growth is frequently limited by water availability. Lixisols often have low levels of organic matter and generally low effective cation exchange capacity (ECEC). They are commonly deficient in phosphorus. Calcisols are extensive, particularly in north and west Africa but, lying on the dry margin of cropland, they contribute less than Luvisols to crop production in semi-arid regions. They are high in bases, in places saline, and low in organic matter. Chemozems are relatively well structured and fertile with high levels of base saturation. Their topsoil does not harden on drying, as is common in Luvisols and Lixisols. Ferralsols and Acrisols are acid, highly weathered and more common on the wetter margins than in strictly semi-arid regions. Being highly weathered, they have low fertility and ECEC though they often have high levels of exchangeable aluminium. Applied fertilizer may be leached, because of the low ECEC, or made unavailable. They have high water infiltration capacity but a relatively low water-holding capacity and are susceptible to compaction.

Vertisols, dark-coloured cracking clays, though not extensive in all continents, are important in some seasonally wet-and-dry cropping systems. They have a high ECEC and base status and are often naturally fertile, though sometimes low in nitrogen, phosphorus and zinc. They have a large water-holding capacity. Their surface does not crust; they self-mulch as surface clods break down naturally on drying to small aggregates forming a good tilth.

The soil cracks deeply on drying, but as it re-wets the cracks close reducing infiltration and lateral movement of water. When wet the soils are so plastic and sticky that, in Australia, Vertisols were not cropped until tractors replaced horses thus enabling them to be ploughed quickly during the short period when soil moisture is optimum. It is useful to draw attention to the major soil types and broad climate and soil groupings, as above. For example, the extensive cereal-fallow-pasture system, covering 60 million ha of temperate semi-arid Australia, is found on four broad types of soil and differences in crop rotations largely reflect regional soil differences. However, the more common situation is that soils are linked through topography and moisture regime to landscape, as illustrated in West Africa.

CLASSIFICATIONS OF CROPPING SYSTEM

Depending on the resources and technology available, different types of cropping systems are adopted on farms. Mono-cropping or Single Cropping: Mono-cropping refers to growing only one crop on a particular land year after year. Or Practice of growing only one crop in a piece of land year after year e.g. growing only rabbi crops in dry lands or only said crops in diary lands (Lands situated in river basins which often remain flooded during rainy season). This is due to climatologically and socio economic conditions or due to specialization of a farmer in growing a particular crop.Groundnut or cotton or sorghum is grown year due to limitation of rainfall. Flue-cured tobacco is grown in Günter (A.P.) due to specialization of a farmer in growing a particular crop. Rice crop is grown, as it is not possible to grow any other crops, in canal irrigated areas, and under water logged conditions.

Monoculture: Practice of repetitive growing only crop irrespective of its intensity as rice-rice-rice in Kerala, West Bengal and Orissa.

Sole Cropping: One crop variety grown alone in pure stand at normal density.

Multiple Cropping or Polycropping: It is a cropping system where two or three crops are gown annually on the same piece of land using high input without affecting basic fertility of the soil.Growing two or more crops on the same piece of land in one calendar year known as multiple cropping. It is the intensification of cropping in time and space dimensions i.e. more number of crops within a year and more number of crops on the same piece of land at any given period. It includes inter-cropping, mixed cropping and sequence cropping.

Molested (1954) has mentioned that multiples cropping is a philosophy of maximum crop production per acre of land with minimum of soil deterioration.

- Rice-potato-green gram.
- Rice-mustard-maize.
- Rice-potato-sesame.
- Jut-rice-potato.

Cropping intensity is more that 200 per cent when the farm as a whole is considered; the Multiple Cropping Index (MCI) is determined by the number of crops and total area planted divided by the total arable area. When the value is three or more, it is said to be most promising farm. This is also called as intensive cropping.

Polyculture: Cultivation of more than two types of crops grown together on a piece of land in a crop season, *e.g.*

- Subabul + Papaya + Pigeon pea + Dinanath grass.
- Mango + Pine apple + Turmeric
- Banana + Marigold + berseem.

Relay Cropping: Growing the succeeding crop when previous crop attend its maturity stage-or-sowing of the next crop immediately after the harvest of the standing crops. Or it is a system of cropping where one crop hands over land to the crop in quick succession, e.g.

- Paddy-lathers
- Paddy-Lucerne.
- Cotton-Berseem.
- Rice-Cauliflower-Onion-summer gourds.

Overlapping Cropping: In this system, the succeeding crop is sown in the standing crop before harvesting. Thus, in this system, one crop is sown before the harvesting of preceding crops. Here the lucre and berseem are broadcasted in standing paddy crop just before they are ready for harvesting.

Advantages:

- Minimum tillage is needed for relay cropping and primary cost of cultivation is less.
- Weed infestation is less, as land is engaged with crops year round.
- Crop residues are added in the soil and thus more organic matter.
- Residual fertilizer of previous crops benefits succeeding crops.

BIOTECHNOLOGICAL CHANGES IN FOOD CROPS

®RICE

It is grown over as much as 29 percent (43,70 million hectare) of the cropped area. India stands next only to China in the production of rice contributing 22.3% of the world production. West Bengal is the largest rice producing state contributing to more than 15.81% of the rice production of India. Next is Andhra Pradesh with 12.75% and Uttar Pradesh with 12.13%. However, yield per hectare is highest in Punjab and Karnataka (42.97% and 31.87% respectively).

WHEAT

Wheat is the second important food grain of India, 13.79% of the gross cropped area is devoted to wheat. Area under wheat increased from 9.7 million hectare in 1950-51 to 26.48 million hectares in 2006-07 and wheat production recorded eleven fold increase. The countries producing more wheat than India are the Soviet Union, The United States and China. Over 35.5 percent of the country's total wheat output comes from Uttar Pradesh, Punjab (20%), Haryana (12.68%). The crop shares about 21 percent for the area under food grains and accounts for about 34 percent of the total food grains production in the country.

JOWAR

Jowar ranks third in area among food grains. The leading jowar producing states are Madhya Pradesh, Maharashtra, Karnataka and Andhra Pradesh. It is both Kharif and Rabi crop and about 8.51 million hectares (5%) are devoted to this crop in India. Half of jowar grow in Maharashtra (51.11%)/

MILLETS

Millets like Jowar, bajra, maize and ragi are important cereal crops of India, They are primarily crops of the dry parts of the Deccan. Gujarat, and rajasthan, These Kharif crops are grown where there is not enough rain to grow rice. They are produced in the order ragi, jowar and bajra, as the rainfall goes on decreasing from the semi-humid to semi-arid regions.

BAJRA

Bajra is generally grown on poor light sandy soils of western Rajasthan (35.94%), Northern Gujarat (13.93%), Uttar Pradesh (16.28%), Haryana (8.85%), and Maharashtra (13.41%).

In the south, it is cultivated on shallow black and red and upland gravelly soils. It occupies 9.48 million hectares (about 5.0% of cropped area of the country).

MAIZE

Maize was introduced in India from America in the beginning of the 17th century. It is an important food crop now in the Great Plains and in the hilly and submontane tracts of the North. Green maize provides sweet and succulent fodder. Maize is an important millet crop and has an area of production next only to jowar and bajra.

It occupies 7.7% of the cropped area of the country. Its highest concentration occurs in Andhra Pradesh (21.1%, Karnataka (18.56%), Bihar (9.25%), Rajasthan (7.48%) and Madhya Pradesh (8.5%) of all India production. Ragi is an important millet, particularly in south Karnataka where millions use it as staple food. It is raise on the red, light black and sandy loams of Karnataka and Tamil Nadu and the well-drained alluvial loams of Uttar Pradesh, Bihar and Gujarat. Barley is an important cereal crop in many parts of northern India. It is also used for melting in the manufacture of beer and whisky.

It is rabi crop in the Great Plains and valleys of the western Himalyas. Uttar Pradesh and Rajashtan together lead in the total area and total production of barley in the country.

PULSES

In our predominantly starchy vegetarian diet pulses form a very important part as they provide us vegetable protein. Being leguminous crops, pulses fix atmospheric nitrogen in the soil and hence are usually rotated with other crops to maintain or restore soil fertility. Pulses are seasonal crops in India. India tops the world in pulses production.

DEFINITION AND SCOPE OF BIOENGINEERED FOODS

The recently proposed rule of the US FDA concerns 'bioengineered foods' which have been defined as 'foods derived from plant varieties that are developed using in vitro manipulations of DNA (generally referred to as rDNA technology)'. As a result, the proposed rule has a much narrower focus than the 1992 US FDA Statement of Policy.

The US FDA has explained the need for a change in emphasis based on their expectations that many of the new plant varieties exhibit a greater potential to 'contain substances that are significantly different from, or that are present in food at a significantly different level than before'. As a result, the substances present in foods and food components derived from new plants developed using rDNA technology are less likely to be considered GRAS, and as a result, will require pre-market approval from the US FDA.

SAFETY AND NUTRITIONAL EVALUATION

The primary objective of the safety and nutritional evaluation is to demonstrate that the food derived from a new plant variety is as safe or nutritious as foods already consumed as a part of the diet. For new plant varieties, including those developed using rDNA technology, a science-based approach is used to focus the evaluation on the demonstrated characteristics of the food or food component.The evaluation of a GM food or food component typically involves reviewing information or data on any newly introduced substances, the known levels of toxicants, as well as the nutritional composition of the plant following modification. Substances that raise safety concerns (*e.g.* toxicants, allergens) would be subject to more extensive evaluation, since both intended and unintended changes may affect the levels of toxicants and nutrients in a food following the modification.

Guidance for performing a safety and nutritional evaluation was provided in the 1992 US FDA Statement of Policy, in a series of flow charts and text that cover:

- The crop that has been modified;
- Source(s) of the introduced genetic material;
- New substances intentionally added to the food as a result of the genetic modification (*e.g.* proteins, but also fatty acids, and carbohydrates).

Documentation required to support the evaluation typically includes: the purpose or intended technical effect of the modification on the plant, together with a description of the various applications or uses, a molecular characterization of the modification including the identities, sources and functions of introduced genetic material; information on the expressed protein products encoded by introduced genes; information relating to the known or suspected allergenicity and toxicity of any expressed gene products; for foods known to cause allergy, information on whether the endogenous allergens have been altered by the genetic modification; information on the compositional and nutritional characteristics of the foods, including anti-nutrients; and in some instances, comparative results of feeding studies involving the foods derived from plants modified using rDNA technology and the non-modified counterpart. In performing its evaluation, the US FDA is particularly interested in the identification of inherent toxicants, known or potential allergens, assessing the concentration and bioavailability of essential nutrients, the safety and nutritional value of any newly introduced proteins, and the identity, composition, and nutritional value of modified carbohydrates, fats and oils. If additional questions of safety remain following this evaluation, further toxicological studies may need to be performed.

It is recognized that absolute assurance of the safety of any food does not exist. As a result, the goal of the safety evaluation is to establish a reasonable certainty of no harm under anticipated conditions of consumption. With this in mind, experience with the existing food supply has provided the basis for evaluating the safety of new food or food components. Both the Food Advisory Committee and Committee for Veterinary Medicine have been extensively involved in the development of approaches for the safety and nutritional evaluation of foods and food components derived from new plant varieties, including those developed using rDNA technology.

PRODUCT CHARACTERIZATION

Product characterization takes into consideration information relating to the modified food crop, the introduced genetic material and its expression product, and acceptable levels of inherent plant toxicants and nutrients. All characteristics of the gene insert must be known, including the source(s), size, number of insertion sites, promoter regions, and marker sequences.

It must be established that the transferred genetic material does not come from a pathogenic source, a known source of allergens, or a known toxicant-producing source. The introduced genetic material should be well characterized to ensure that the introduced gene sequences do not encode harmful substances and are stably inserted within the plant genome to minimize any potential opportunity for undesired genetic rearrangement. Analytical data are required to evaluate the nutritional composition, the levels of any known toxicants, antinutritional and allergenic substances, and the safety-in-use of antibiotic resistance marker genes.Any new substances introduced into crops through rDNA technology (*e.g.* proteins, fatty acids, carbohydrates) will be subject to pre-market review as food additives by the US FDA, unless substantially similar to substances already safely consumed as a part of foods, or that are considered GRAS. To date, substances that have been added to foods through rDNA technology have been previously consumed or have been determined to be substantially similar to substances already consumed as a part of the diet. As such, introduced substances have been considered exempt from the requirement for pre-market approval as food additives with the US FDA. A more rigorous safety evaluation of a GM crop is warranted if the introduced gene sequence(s) has not been fully characterized, the nutritional composition has been significantly altered, antibiotic resistance marker genes have been used during its development, or if an allergenic protein or toxicant has been detected at levels higher than what is typically observed in edible varieties of the same crop species. In any event, determinations as to the safety of substances that have been introduced into new plant varieties through rDNA technology are made on a case-by-case basis.

Although an evaluation of the introduced gene sequences and expression product(s) provides assurance as to their safety, further studies may be required to predict whether unexpected effects may result following their interaction with other genes within the plant. In addition, the product characterization of a GM plant involves assessing sequence homology to known toxicants and allergens, thermal and digestive stability, and if required, the results of both in vitro and in vivo assays to demonstrate lack of toxicity.

COMPOSITIONAL ANALYSIS

The results of field trials performed over several years serve to characterize the phenotypic and agronomic characteristics exhibited by the plant (*e.g.* height, colour, leaf orientation, susceptibility to disease, root strength, vigour, fruit or grain size, yield, etc.), as well as to provide the materials required for the compositional analysis. Any anomalies in the phenotypic or agronomic characteristics exhibited by a plant may result in a requirement for additional information. Protein, fat, fibre, starch, amino acid, fatty acid, ash and sugar levels are determined, as well as the levels of anti-nutrients, natural toxicants or known allergens.

Studies of the nutritional composition are performed to determine whether the levels of any key nutrients, vitamins or minerals have been altered as a result of the genetic modification. Based on the results of these studies, a determination is made as to whether the phenotypic and agronomic characteristics of a GM crop or the concentrations of inherent constituents fall within ranges typical of its conventional counterpart.

If inserting a new gene causes no change in any of the assessed parameters, the US FDA can conclude with reasonable assurance that the GM crop is as safe as the conventional crop. If the levels of essential nutrients or inherent toxicants are found to be significantly different in the GM crop, the US FDA may recommend additional action prior to commercialization, such as obtaining food additive status, or the use of specific labels to alert consumers of an altered nutritional content, etc.

ALLERGENICITY

In consultation with scientific experts in the areas of food safety, food allergy, immunology, biotechnology and diagnostics, the US FDA published guidelines for assessing the allergenicity of GM foods or food components in 1994. The approach to assessment is multi-faceted, incorporating data regarding the origin of the genetic material, and the biochemical, immunological and physicochemical properties of the expressed protein. The overall assessment is reliant upon the fact that all known food allergens are proteins and, notwithstanding the number of shared properties between allergenic and non-allergenic food proteins, food allergens tend to exhibit a number of similar characteristics. In general, food allergens share a number of common properties: they have a molecular weight of over 10 000 Da; they represent more than 1% of the total protein content of the food; they demonstrate resistance to heat, acid treatment, proteolysis and digestion; and they are recognized by IgE.

For gene sequences derived from known allergenic sources (*e.g.* peanuts), the developers of GM plants are expected to demonstrate that allergenic proteins have not been introduced into the food. For assessment purposes, it is assumed that any genetic material derived from a known allergenic source will encode for an allergen. To demonstrate otherwise, the amino acid sequence of an expressed protein must be compared with that of known allergens using protein sequence databases.

Furthermore, in vitro and/or in vivo immunologic analyses using the sera of allergic patients sensitive to the source of the genetic material may need to be performed to determine whether or not a potentially allergenic protein is being expressed in the GM food. Some GM foods may be modified to express genes from a source that is not known to be allergenic when consumed.

Under these circumstances, the US FDA follows a similar decision tree-based approach to determining the allergenic potential of the expression product. In assessing these proteins, should any amino acid sequence exhibit homology with a known allergen, the expressed protein would then be evaluated in immunologic tests using the sera of patients known to be allergic to the identified homologous protein. Regardless of the origin of the genetic material, physicochemical studies are performed in vitro to provide information concerning the expected stability of the expressed protein.

All known food allergens tend to be resistant to digestive degradation, as demonstrated in simulated gastric fluid models, or to decomposition under conditions of food processing. In making a determination regarding a GM food, it is the totality of the biochemical, immunological and physicochemical properties of the introduced protein that provides guidance as to the allergenic potential of such a protein being expressed in food.

The level of protein expressed, produced and consumed as a part of the diet is also a primary indicator of the allergenic potential, since nearly all food allergens are known to be major proteins in their respective foods. If the results of any of these studies suggest an allergenic potential, the US FDA may recommend further scientific evaluation, require special labelling to alert sensitive consumers, or alternatively caution the developer about proceeding with the development of a particular GM food. Most recently, a joint FAO/WHO Expert Consultation on Allergenicity of Foods Derived from Biotechnology recommended a revised decision tree. This decision-tree strategy was modified from the previous version FAO/WHO to include a revised definition of sequence homology for gene product comparison; a greater emphasis on serum testing, even with gene products without homology to known allergens and not derived from an allergenic source; and animal models to assess potential allergenicity, despite acknowledgement by the Expert Consultation that these models are currently under development and at present, not predictive of food allergies in humans.

More recently, the ad-hoc Open-Ended Working Group on Allergenicity established by the ad-hoc Intergovernmental Codex Task Force on Foods Derived from Biotechnology, considered the FAO/WHO strategy in drafting an approach to assessing the potential allergenicity of foods derived from rDNA plants.

The Codex Working Group recognized the absence of a definitive predicitve test for allergenicity in humans to a newly expressed protein and recommended an integrated, stepwise assessment strategy.The strategy recommended by the Working Group, and accepted by the Codex Task Force for inclusion in the draft forwarded for final adoption by the Codex Alimenarius Commission, is consistent with that of FAO/WHO. However, the Working Group suggested that some of the modifications included in FAO/WHO could contribute to the overall weight of evidence of any conclusion of potential allergenicity (*e.g.*, allergen specific serum depositories, animal models), pending development and validation.

ANTIBIOTIC RESISTANCE

The use of antibiotic-resistance genes as selectable markers has been common practice in the development of new plant varieties using rDNA technology. Concerns relate to the potential transfer of antibiotic-resistance genes from GM plants to pathogens in the environment or to the gut of humans consuming foods or food components. Issues relating to the use of antibiotic resistance genes were identified in the 1992 US FDA Statement of Policy Statement as well as discussed in additional guidance entitled Guidance for Industry: Use of Antibiotic Resistance Marker Genes in Transgenic Plants.

The guidance provided within these documents was established in consultation with experts in the fields of microbiology, medicine, food safety, bacterial and mycotic diseases, and includes suggestions with respect to the continued safe use of antibiotic-resistance marker genes by the developers of new plant varieties. The use of marker genes that encode resistance to clinically important antibiotics has raised questions as to whether their presence in food could reduce the effectiveness of oral doses of the antibiotic or whether the gene present in the DNA could be transferred to pathogenic microbes, rendering them resistant to treatment with the antibiotic.

The risk of transfer of antibiotic-resistance genes from plants to microorganisms considered to be pathogenic to humans, however, is considered to be minimal if not insignificant. Furthermore, the potential risks are becoming less of a concern as more developers are beginning to research the use of alternative technologies (*e.g.* non-resistance-based markers) in plant breeding. The conclusions with respect to the safe use of antibiotic-resistance marker genes are consistent with the findings of other national and international food safety organizations.

Consultation and Filing Process

The submission of a Pre-market Biotechnology Notification (PBN) has recently been proposed as a mandatory requirement for the commercialization of bioengineered foods and food ingredients within the United States. Minimal differences exist between the information to be submitted as a part of a PBN and that previously presented in voluntary consultations with the US FDA.

As proposed, developers are still being encouraged to consult with the US FDA as early and as often as necessary in the development of a bioengineered food, such that any potential scientific or regulatory concerns can be identified and addressed prior to the submission of the PBN. Guidelines for performing consultations with the US FDA were released in a publication entitled Guidance On Consultation Procedures: Foods Derived from New Plant Varieties in 1997.

The publication recommended an approach for developers to proceed by submitting a request for consultation, outlined the internal process by which all requests would be handled, and included additional guidance as to the type of safety and nutritional information to be presented during consultation with the US FDA. Once sufficient safety and nutritional information has accumulated to demonstrate that a product is safe and in compliance with the FFDCA, developers typically schedule a consultation to present their scientific findings and conclusions to the US FDA. Consultations prior to notification not only serve to keep the US FDA informed of advances made in the application of rDNA technology in food production, but also keep the developers of bioengineered foods aware of emerging safety, nutritional or regulatory concerns of the US FDA. Consultations are considered complete when all safety and regulatory concerns between the US FDA and the developer have been resolved. The US FDA proposes to perform an initial evaluation of a PBN within 15 days of receipt to determine completeness, at which point, if considered complete, the PBN will be filed, and a response can be expected within 120 days.

The US FDA does not issue a product approval per se, but informs the developer by letter that:

- The evaluation period has been extended;
- The notice is not complete and why;
- It has no further questions 'at this time' based on the information that has been presented.

Labelling

The FFDCA defines what information must be disclosed to consumers on a food label, such as the common or usual name, and other limitations concerning the representations or claims that can be made or suggested about a food product. All foods must be labelled truthfully and not be misleading to consumers. Taking this into consideration, the FFDCA does not stipulate the disclosure of information on the basis of consumer desire to know. Labelling may be considered misleading if it fails to reveal material facts in light of representations that are made with respect to a product. The labelling of foods derived from new plant varieties, including plants developed using rDNA technology, was originally addressed in the 1992 US FDA Statement of Policy, and most recently discussed in Draft Guidance for Industry for the voluntary labelling of bioengineered foods.

To date, the US FDA is not aware of any information that would distinguish foods developed using rDNA technology (*e.g.* bioengineered foods) as a class from foods developed through other methods of conventional plant breeding, and as such, have not considered the method of development a material fact requiring disclosure on product labels. Nevertheless, after extensive consultation, including thousands of written comments and a series of public meetings, the US FDA has observed 'a general agreement that providing more information to consumers about bioengineered foods would be useful'.

Requirements for Labelling

Special labelling is required if the composition of the bioengineered food differs significantly from its conventional counterpart. For example, for a food that has been genetically modified to contain a new major sweetener, a new common or usual name or other labelling may be required.

Similarly, if a GM food contains an allergen that consumers would not expect to be present in that food, special labelling may be necessary to alert sensitive consumers. If a protein commonly associated with an allergic reaction (*e.g.* peanut protein) is transferred to another food through genetic modification, the US FDA would evaluate whether labelling would provide sufficient consumer protection. If labelling would not be considered to provide a sufficient level of protection, the US FDA would take appropriate steps to ensure the GM food would not be marketed.

Therefore, current policy requires a GM food to be labelled when the resulting product poses a safety issue or is substantially different from its conventional counterpart, and as a result, could be considered to pose a misrepresentation to consumers. The 1992 US FDA Statement of Policy and the recent Draft Guidelines do not consider the use of rDNA technology in the development of food products to be a material fact requiring specific disclosure on the label. Rather it is a method of development, similar to other methods of plant breeding, which have not required disclosure on the label. Bioengineered foods cannot be distinguished compositionally from foods modified through more conventional methods and thus do not require specific disclosure through labelling. The Draft Guidelines reaffirm the US FDA position that bioengineered foods do not require special labelling.

Voluntary Labelling

To provide guiding principles for voluntary labelling, in recognition of the desire of certain manufacturers to label foods as produced either with or without bioengineering, the US FDA published a 'Draft Guidance for Industry: voluntary labelling indicating whether the foods have or have not been developed using bioengineering'. Emphasizing that the use of rDNA technology 'is not a material fact', the US FDA recognizes that some consumers want disclosure of bioengineered content and that some manufacturers wish to provide it. In response, Draft Guidance was issued with suggestions concerning the use of labelling statements that are not considered misleading.

PRESENCE OR USE OF RDNA

The US FDA provided several examples of how disclosure of bioengineering can be accomplished, be informative and not be misleading.

- Example 1 'Genetically engineered' or 'This product contains corn meal that was produced using biotechnology'. These disclosures reveal the minimum amount of optional information about bioengineering.

- Example 2 'This product contains high oleic acid soybean oil from soybeans developed using biotechnology to decrease the amount of saturated fat.' This statement explains the proper and required name of this type of soybean oil, since it differs from what would be considered as soybean oil. The optional comments about biotechnology and decreasing saturated fat provide information that could be seen as benefits of the product.
- Example 3 'These tomatoes were genetically engineered to improve texture.' This example was one included to illustrate how it could be misleading to consumers if they cannot discern a difference in texture, but would not be misleading if they can tell a difference. If the former, and the new texture is to facilitate processing, then this intention should be made clear, *i.e.* 'to improve texture for processing'.
- Example 4 'Some of our growers plant tomato seeds that were developed through biotechnology to increase crop yield.' This is another example of optional information that would explain an indirect, agricultural benefit.

ABSENCE OR NOT BIOENGINEERED

The US FDA provides an important commentary in the Draft Guidance about 'genetically modified organisms (GMO)' versus 'bioengineered'. It would be technically inaccurate to use the phrase 'not genetically modified' or 'GMO-free' to mean that bioengineering was not used.

This is because most conventional foods have been genetically modified over the years by traditional crop breeding practices.

Examples of acceptable voluntary statements would include:

- We do not use ingredients that were produced by biotechnology.'
- This oil is made from soybeans that were not genetically engineered.'
- Our tomato growers do not plant seeds developed using biotechnology.'

Another important point is made about a term such as 'GMO-free' that is misleading for two reasons:

- Many foods do not contain organisms anyway and therefore should not be labelled as 'organism-free';
- Free' implies complete absence or 'zero' amount of bioengineered material. In a practical sense, it is impossible to demonstrate analytically the complete absence of anything. Therefore, in reality, a threshold for possible adventitious presence of a low level of bioengineered material may be necessary and has been the subject of much debate.

The Agency also provides additional guidance about misleading statements that:

- Could be interpreted to suggest that the absence of bioengineering would make the food superior to the bioengineered alternative;
- Claim the absence of one bioengineered ingredient when the food contains another ingredient that is bioengineered;
- Claim that a food is not bioengineered when in fact this type of food (*e.g.* green beans) has never been modified through rDNA technology.

GENETIC TRANSFORMATION IN PLANT BREEDING

The application of recombinant DNA technology has revolutionized plant breeding in recent years. Genetic transformation in plant breeding greatly increases the gene pool and broadens the scope of genetic changes that modern breeders may draw upon. The first generation of genetically modified (GM) crop plants hold great promise in carrying new or improved agronomical traits that require less intensive farming methods or which have improved quality traits.

These GM crop plants have been produced to improve farming production systems, aiming, for instance, to reduce negative impacts on the environment, such as fertilizers, herbicides and pesticides. In general they are of little direct interest to the consumer. However, the genetic manipulation of the primary and secondary metabolism of plants offers distinct possibilities in the development of nutritionally improved products, raw materials with added pharmaceutical value, and crops with real health benefits for the consumer.

These are second-generation GM plants. Recently, several prototypes in molecular plant breeding with novel traits for (non-) food and industrial applications have been introduced into the market, and many more GM crops have the prospect of commercial exploitation in the near future. The safety and wholesomeness of these GM crop plants raises questions beyond those posed by conventional foods. In this context, traditional plant breeding of most food crops has conformed to the standard 'generally recognized as safe' (GRAS). In some instances safety is assumed to be assured during performance and quality testing by breeders (*e.g.* glucosinolates and erucic acid in rape seed).

However, the process is largely focused on the comparison of agronomic and phenotypic characteristics, such as colour, maturity, yield or disease resistance. Products with an unusual taste or that have harmful effects following ingestion or skin contact have normally been rejected from the breeding programme. However, the safety of conventional plant varieties has usually been presumed from the assumption that prudent consumers will avoid those species once it is known that they cannot consume them without adverse health effects. When the commercial production of GM plants started, it sparked a debate centred on issues associated with concerns on environmental, human and animal safety matters. Many people questioned whether the 'simple' GRAS approach, based on historical and empirical experiences, is sufficient to guarantee food safety of the GM crop plants. It is also asked whether the risks of chronic exposure of humans and animals to GM crops are adequately covered, and whether large-scale breeding of GM plants will have detrimental effects on ecosystems.

VARIETY REGISTRATION OF EXISTING CROP PLANTS

Although the variety registration of existing crop plants has not resulted in adverse effects in humans, the conventional assessments carried out by plant breeders was not considered to be sufficient to ensure food safety of GM crop plants. As a result, Europe issued a regulation on novel foods and novel food ingredients that sets the legal framework for the market introduction of genetically modified organisms (GMOs). This came into force. The Novel Foods Regulation establishes a system of mandatory pre-market approvals of novel foodstuffs, including GM plants.

The accompanying EU guideline gave a clear indication of the types of data that are needed to form the basis of an assessment. In particular, four central safety issues have been set forth related to food and feed safety:

- The nutritional and toxicological consequences of inserted gene products;
- The potential of pleiotropic (unintended) effects in the host organism due to the insertion event;
- The allergenicity of expressed proteins and novel foodstuffs;
- The potential of gene transfer to human and animal gut flora.

This means that in transgenic insect-resistant Bt tomatoes encoding the C-terminal truncated Bt2 gene derived from a Bacillus thuringiensis (Bt) strain, IAb5, the novel protein has to be evaluated on its own merit, while the safety and wholesomeness of the 'remaining' transgenic tomato is assessed separately. Whereas, for example, in antisense RNA exogalactanase tomatoes, which have improved rheologic characteristics due to downregulation of the endogenous exogalactanase activity, the safety assessment is primarily focused on the 'remaining' novel fruit. In general, there is basic agreement on the safety issues to be addressed. One particular area of concern is the possibility of unexpected or unintended metabolic perturbations due to genetic modification that may alter, for instance, levels of nutrients and health-influencing components. The use of rDNA techniques does not necessarily result in fundamental changes in crop plants compared to the food produced by conventional breeding methods. However, it should be emphasized that a uniform international agreement for the evaluation of GMOs is urgently needed.

For example, an agreement on how to establish in practical terms the similarities or differences between the 'remaining' GM crop plant and the traditionally bred plant seems to be much more difficult to achieve. This chapter is devoted to the description of advanced strategies for the evaluation of potential unintended effects in GM crop plants. Pleiotropic (unintended) alterations in agronomic traits or composition may arise from insertion mutagenesis or as a result of metabolic effects of the novel gene product(s).

The challenge is to gain the ability to discriminate between metabolic alterations due to somaclonal variations, unintended effects or natural diversity. Thereto, emphasis is given to a platform of technologies designed to identify changes in the molecular machinery of crop plants. As is the case for detection methods for GM crops and derived products, there are needs and criteria for sampling strategies, statistical models and manufacturing standard reference matehers recently.

UNINTENDED EFFECTS IN FOOD SAFETY AND WHOLESOMENESS

Strategies for assessing the food safety and wholesomeness of novel foodstuffs are currently in the exploration phase. Above all there is a need for uniform and harmonized quantitative methods for the testing of potential unintended effects in whole foods, such as GM crop plants are. The safety of, for example, synthetic chemicals and food additives is established by assessing separate elements of the compound in question.

Although very successful in case of single chemicals, the evaluation of a whole food product may not be a simple addition of individual assessments of the constituents. Traditional testing protocols involving labouratory animal 90-day feeding trials with (whole) food products are far from ideal. This type of animal feeding study with complex food matrices is complicated by the likelihood of nutritional imbalances leading to dietary problems, confounding factors, an insensitivity for specific endpoints, and the impossibility of using large safety margins, if any. Therefore, in a global context several regulatory, research and industry-based bodies have proposed alternative concepts for the safety evaluation of genetically engineered foods and food ingredients. In particular the Organisation for Economic Co-operation and Development (OECD), the Food and Agriculture Organization (FAO) and the World Health Organization (WHO) have developed concepts and principles for the safety evaluation of GMOs and derived products.Their reports concluded that risk assessments should be directed at demonstrating that a GM crop plant or derived food product is as safe as a traditional or previously authorized product. Accordingly the EU Competent Authority incorporated these general principles into their Novel Foods Regulation and accompanying guidelines. In this context, novel foods and food ingredients that are considered to be 'no longer equivalent' to traditionally bred plants are subject to labelling as defined in the Council Regulation.

SUBSTANTIAL EQUIVALENCE

The term 'substantial equivalence' appears in the EU Novel Foods Regulation as a major principle in the risk assessment. It emphasizes that, within the context of the Regulation the evaluation of the safety and nutritional value of a GM plant should be focused on a comparative analysis with conventionally bred products. OECD's Group of National Experts on Safety in Biotechnology enunciated this comparative approach towards the safety assessment of transgenic food plants as the concept of substantial equivalence. In their report, it is assumed that existing food organisms possess a long history of safe use (*i.e.* GRAS). Consequently, the most practical approach to the determination of safety is to consider whether the GM plants are substantially equivalent to analogous food product(s), if such exist. Once substantial equivalence of a novel foodstuff has been established, it provides assurance of safety that is equal to or better than that of its comparators.

Further clarification of the concept of substantial equivalence originates from a WHO sponsored workshop. It has been concluded that the establishment of substantial equivalence is not a safety assessment per se, but a dynamic, analytical exercise. Thus, it may mean that analysis of, for instance, gene expression patterns, global changes in protein expression and/or differences in metabolic capabilities (the metabolome) of the novel product in comparison with conventionally bred products should be performed in order to examine equivalence.

This strategy has worked satisfactorily for the assessment of the first-generation crop plants for which sufficient background knowledge is available. However, relatively less experience has been gained with the safety and wholesomeness evaluation of novel food plants for which no substantial equivalence can be established. Even in the case of single gene modification, potential alterations in metabolic pathways are difficult to identify and, consequently, the implications of the genetic modification process on the metabolism of plants are poorly understood. Moreover, data on the mechanisms by which plants regulate, for instance, their response to environmental stress factors and other conditions are remarkably scanty.

Critical Nutrients and key Toxicants

Application of the concept of substantial equivalence appears in practice to lead to various interpretations. Different requirements for risk assessment with regards to the potential of unintended effects exist as a result of genetic modification. Actual approaches entail a consideration of the characteristics of host and donor organism, phenotypic properties and toxicological evaluation and include comparative determinations of any changes in critical (anti-)nutrients and key toxicants for the food source in question.

In general, the analysis of an expanded spectrum of components is unnecessary, but should be considered if there is an indication from other traits that there may be an unintended secondary effect of genetic modification. However, compositional analyses based on single parameters using a list of crop-specific critical (anti-)nutrients and key toxicants as a minimum requirement of screening for possible variations in the content has its limitations.

Limited data are typically available for the less important food components, as in less well-known crops most critical (anti-)nutrients and natural toxicants will be unknown. Moreover, the toxicants to be assessed may be partly influenced by the cellular function of the product encoded by the inserted gene. Furthermore, the natural variation of critical comparators may mask the evidence of secondary effects due to genetic modification.

Thus, there is no criterion that determines whether a difference between a GM plant and its natural counterpart is outside the literature values for the ranges of parameters that express natural diversity. Relevant information on indicators of unexpected effects may not have been observed during the development of an edible transgenic crop species. In the future other compounds could become of importance, such as anti-oxidative, oestrogenic and anti-carcinogenic compounds.

In determining critical nutrients and/or key toxicants, differences in consumption patterns and practices of processing and consumption in various geographical settings and cultures must also be recognized. It is our opinion that conclusions about relative safety and wholesomeness based on a list of selected single components may not be equally valid in a global context and in all times. Therefore, there is an urgent need to set up generic approaches using platform technologies to evaluate undesirable metabolic perturbations in GM food crops.

A Post-genome Challenge

In our view the identification of unintended effects encompasses the ability to interpret and use innovations in molecular genetics research such as genomics, proteomics and metabolomics on a crop-by-crop basis. This will establish a whole data package directed towards a holistic view on possible unintended side effects due to genetic modification. In essence, the focus is shifting towards molecular characterization to understand functional activity.

Vital to this approach will be informative profiles of molecules at different integration levels, *e.g.* mRNAs, proteins including post-translational modifications and molecules at the level of primary and secondary plant products and metabolites that might be useful benchmarks for the detection of unintended effects. The relative importance of these various 'profiling' techniques in establishing data sets for assessing unintended effects is not indicated by their order and will vary from species to species. Alterations in, for example, expression levels, post-transitional modification, interactions or chemical composition do not per se imply that the product is less safe or even unsafe. Such a platform of technologies may determine whether unintended effects exist between the GM plant and its control(s) or whether there is substantial equivalence apart from certain well-defined expected differences. Measures of the relative activities of various molecular constituents associated with metabolic capabilities in non- and modified plant cells under different conditions will provide new insights into how metabolism and genetic modification are orchestrated. But it should be emphasized that the fields of plant genomics and proteomics are still in their infancy.

EQUIVALENCE OF GENETICALLY MODIFIED CROPS

A very promising strategy to evaluate the equivalence of genetically modified crops is gene expression profiling. Conventional methods for the analysis of differential gene expression include Northern blotting, S1 nuclease protection, comparative expressed sequence tag (EST) sequencing, differential display and serial analysis of gene expression. More recently, Kok *et al*. have designed a method of detecting altered gene expression by means of mRNA fingerprinting or reverse transcription polymerase chain reaction (RT-PCR). Based on the original concept of Liang and Pardee specific subsets of the mRNA population of a tomato plant were amplified. This enables the expression of different genes in the GM crop and genes in the parental line to be compared and any empirical information about significant alterations to be established.

The method is still under development and requires further validation, but has been shown to be reproducible and able to detect important differences in gene expression. Some of the most significant remaining problems are that the method largely depends on individual skills based on experience and is time-consuming, even in its most simplified form.

CDNA MICROARRAYS

Over the past few years, the DNA microarray technology has emerged as a powerful high-throughput method for the analysis of gene expression. At present, microarrays for gene expression studies generally consist of cDNAs, which are physically deposited onto small glass surfaces. The major advantage of DNA microarray technology over conventional gene-profiling techniques is that it allows small-scale analysis of expression of a large number of genes in a sensitive and quantitative manner. Furthermore, it allows comparison of gene expression profiles under a large number of different conditions. The cDNA microarray technology has already been successfully introduced in disciplines ranging from cell and developmental biology to drug development and pharmacogenomics. In the following a number of technical aspects of the technology as well as its potential value for the safety assessment of genetic modification of food plants.

The cDNA microarrays are produced by deposition onto a solid surface of purified PCR products corresponding to specific genes. Typically a few nanolitres of DNA solution (100-500 μg/ml) are spotted using a micro-dispensing robot. Essential information on robot construction and protocols can be found on the Internet. In general, with respect to the spotting devices, a higher density of spots is accomplished when using a passive dispenser (> 2500 DNA spots/ cm2).Thereto, glass slides covered with a positively charged layer (*e.g.* poly-l-lysine) or carrying reactive groups are most commonly used. To analyse the expression of genes in a plant cell or tissue sample, polyA+ RNA (typically 0.5-2.5 μg is required for each reaction) is purified and labelled by reverse transcription using an oligo-dT primer incorporating a fluorescently labelled nucleotide. The labelled cDNA pool is then hybridized to the microarray. The required hybridization conditions (*i.e.* sample concentration, stringency of hybridization) and parameters, such as detection level and the correlation between transcript concentration and hybridization have been determined for each experimental setting.

The intensity of a hybridization signal, which is a measure of the relative expression level of each gene, can be read by charge-coupled device (CCD) cameras or confocal/non-confocal laser scanners. RNAs extracted from two different samples can also be simultaneously analysed with a microarray by labelling each with a different fluorescent label using either Cy-3 or Cy-5. At RIKILT-DLO we are currently testing whether the analysis of differential gene expression using DNA microarrays is an informative strategy for the safety evaluation of GM plants.

To address this question, DNA libraries are constructed and enriched by a subtractive cloning approach for cDNAs preferentially expressed in either green or red tomato fruit. These subtracted cDNAs, as well as control cDNAs representing known tomato genes (*i.e.* sequence information derived from public sequence databases), have been spotted on arrays and are used for comparison of gene expression profiling of control and genetically modified tomatoes.In the near future we will construct more defined arrays containing genes of toxicologically relevant pathways, such as those involved in the synthesis of natural plant toxins (*e.g.* the glycoalkaloids d-tomatine and d-solanine).

CHEMICAL FINGERPRINTING

The principles for establishing a chemical fingerprint were directed towards the detection of alterations in compositions in 1H-NMR spectra arriving from two populations, for example, non-modified counterpart(s) and an engineered one. In this way the unprocessed parts of the plant used for human consumption were fractionated and analysed, *i.e.* in tomato comparison on the fruit level.

Genetically modified tomato varieties, such as the antisense RNA exogalactanase fruit, have been studied using this innovative technology. The spread of measurements of the individual profiles was comparable for the GMO and its controls. By subtraction of the 1H-NMR spectra the differences of the means were calculated at 99% confidence intervals between the GM crop average (n = 8/batch per line) and its isogenic control average (n = 8/batch per line). The difference of the means was defined here as all differences of the means of the amplitudes of individual NMR signals obtained from spectra of various populations. The differences of the means were used to quantify how the transgenic crop plant differed from its parental line. The normalized 99% confidence intervals of the normalized means of fractions A-E after eightfold independent replicates demonstrated that differences in amplitudes exceeding at least 20% could normally be detected. This is an adequate sensitivity to allow statistical evaluation, and is in line with recommendations of the Nordic Council. They proposed that if the average value of a parameter differs by more than 20% an explanation should be sought.

With the purpose of establishing those differences that may arise from metabolic effects linked to genetic modification the default statistical analysis used a mixed effects randomized block model. Random block effects such as variations in the stage of cultivation, location, logistics or climate had a significant impact on the overall chemical fingerprints of cultivars. For example, chemical fingerprints varied much more between field sites and seasons than between replicates in one plot.

A tomato background mean (*i.e.* crop mean value) was established by constructing an extended range of non-modified tomato references for comparison with natural variability, and to assess the impact of external factors (*i.e.* variations in season, climate, etc.). The crop mean value was used to quantify how the differences of the means between the transgenic crop plant and its control differed from the natural diversity of commercial counterparts.

To differentiate between compositional changes either due to genetic modification, genetic variability or environmental variations the chemical fingerprints were studied in a hierarchical approach.

Hence, the fingerprints of the GMO are compared to those of:

- The isogenic control line grown side-by-side under identical conditions in one plot;
- The isogenic control line grown side-by-side under identical conditions at multiple sites;
- An extended range of commercial varieties of that crop;
- The influence of downstream processing.

It was recognized that 45% of all compounds of the control line harvested in different years varied considerably in concentration. Most differences were found in fractions A, C and D. A similar degree of impact on composition was observed in the case of processing temperatures (*e.g.* 100°C versus 70°C) on tomato juice. These findings were compatible with the protease activity surviving the 70°C treatment. But in fruit harvested from various individual plants, divided into two populations, approximately 95% of all constituents were practically identical in concentration.

Also the determination of the differences of the means between transformant tEG to its isogenic control bred side-by-side in the same plot showed that 249 out of 3000 amplitudes (8%) varied in concentration. Assignments by NMR indicated that, for example, citric acid was present in fraction A of the control at 1.4 times the levels found in the transformant. However, in fraction A of the transformant glutamic acid and/or glutamine showed an increase of 1.3 times the levels found in the isogenic control.

On the other hand, one single aromatic compound, d-lycopene, appeared to be present in fraction E of transformant tEG at 2.5 times the concentrations found in the parental line. No alterations were observed in ß-carotene levels. These effects may be the result of the antisense downregulation of the tEG1A gene in tomato. In our view the natural diversity must be taken into consideration, however, when interpreting the biological relevance of any significant difference found in the differences of the means between the GMO and its parental line. The parental line was extended by entering non-isogenic commercial references (*i.e.* crop mean value).

Although limited in the range of lines, the comparison of the crop mean value to the differences of the means showed that the contribution and magnitude of differences (*e.g.* citric acid, glutamic acid/glutamine, d-lycopene) diluted out. It was recognized that in the agricultural practice the conditions of culturing, processing and storage of appropriate references will be mostly unknown and they may have a non-isogenic genotype. Notwithstanding these complicating random block effects we recommend that, for example, the chemical fingerprint should not be limited to the transgene and its isogenic parent bred under identical circumstances only, but should also include a sampling through multiple sites and a comparison with an extended range of controls of that crop.Otherwise it would be a too limited approach towards the establishment of the claim of substantial equivalence and it would neglect the ranges of natural variations in that crop. In future, if many more lines have to be compared with each other this approach may benefit from multivariate statistical methods.

PROTEOME PROFILING

The introduction of Bt genes of bacterial origin may theoretically influence the nature of post-translation modification (*e.g.* glycosylation) in the transgenic tomato fruit. This is because as well as the impact of the protein moiety, for example, the type of glycosylation pattern may alter due to inactivation of glycosidases and glycosyltransferases in the host cell. In addition, glycosylation of the inserted Bt gene product in the plant might result in altered protein activity and stability.

The importance of structurally elucidating these oligosaccharides, especially the N-glycans, can be attributed to their widespread functions in the cell. These functions include correct folding, biological activity and stability of proteins as well as involvement in plant development. Moreover, N-glycans represent antigenic epitopes by themselves. Immunological as well as biochemical studies revealed the allergenic potency for the d1,3-fucosylation of the proximal N-acetylglucosamine residue. Additionally, also the e1,2-xylosylation of the e core mannose is suspected to be an allergenic epitope. Recently, there is increasing evidence that these N-linked oligosaccharides may have a substantial impact on the immunogenicity of glycoproteins, as N-glycans are suspected to play a certain role in the adverse food reactions of hypersensitized patients. At the Institute for Agrobiotechnology, Austria, this issue was tackled by developing a generic technology for isolating, purifying and characterizing N-linked oligosaccharides in plant tissues. Our low-temperature acetone powder method appeared to be most suitable for the isolation of (glyco-)proteins from plant tissue (*e.g.* tomato fruit). The resulting (glyco-)protein fraction was proteolytically digested and corresponding glycopeptides purified by cation-exchange chromatography.

The oligosaccharides (N-glycans including d1,3-fucosylated oligosaccharides) were released from the peptide backbones by N-glycosidase A (PNGase A) and fluorescently labelled with 2-aminopyridine. Subsequently, pyridylaminated glycans were separated and analysed structurally by repeated two-dimensional high-performance liquid chromatography (2D HPLC) using a size-fractionation and a reverse-phase column in combination with an exoglycosidase treatment.

MALDI-TOF (matrix-assisted laser desorption ionization time-of-light) mass spectrometry was applied for further confirmation of the observed structures. Sixteen different N-glycosidic structures could be detected in tomato fruit. The two most abundant glycans in all the tomato samples showed identical properties to those of the major N-linked oligosaccharides of horseradish peroxidase (MMXF: Man(3)Xyl-FucGlcNAc(2)) and pineapple stem bromelain (MOXF: Man(2)Xyl-FucGlcNAc(2)), respectively, and accounted for about 65-78% of the total glycan content.

Oligomannosidic glycans occurred in only small quantities (3-9%). The majority of the N-glycans found was e 1,2-xylosylated and carried an d1,3-fucose residue linked to terminal N-acetylglucosamine. A structural element was identified and shown to be an IgE-reactive determinant, which contributes to cross-reactivities among non-related glycoproteins, and is present in a wide variety of plants extracts.

It is proposed that this structural element might also be of importance in some cases of food and pollen allergy. The carbohydrate profiling of, for example, populations of green and red-ripe fruit of non-modified and modified Bt tomato plants (expressing the cryIA5 gene) revealed no differences in the nature of N-glycans, nor were there significant differences in the relative amounts of glycan.

Interestingly, the relative amounts found were quite similar across seasons and comparable to the commercial variety Notoro used as an extended control. As far as the N-linked oligosaccharides were involved it was not possible to detect significant alterations in the d1,3-fucose content of Bt tomatoes due to genetic modification.

PROFILING OF THE METABOLOME

Among other factors, identity of a complex GM crop plant with its traditionally bred parent entails metabolism and the level of (un-)desirable substances contained therein. Therefore, we would like to build up an overall picture of components associated with different metabolic capabilities in the (non-)modified plant tissue. A multi-compositional analysis of the so-called metabolome may overcome the test limitations of, for example, single critical nutrient or key toxicant analyses and problems related to their limits of natural diversity.

An attractive alternative is the utility of high-resolution proton (1H) nuclear magnetic resonance (NMR) spectroscopy (1H-NMR) in combination with separation techniques (liquid chromatography (LC)). Recently, we have shown that the approach of chemical fingerprinting using off-line LC-NMR can obtain information on possible changes in complex plant matrices due to environmental effects as long as GLP-like conditions for sample handling, data acquisition and automation of data handling are maintained. In principle, it will be possible to screen all the different low molecular weight components (MW <10 kDa) present in the metabolome of the plant tissue.

TOXICOLOGICAL PROFILING

The significance of detected compositional alterations may need to be further explored by in vitro toxicity assays in order to screen for potential adverse effects and for mechanistic studies. It may even be that the latter approach leads to a redesign, refinement or even replacement of (sub-)chronic rodent feeding trials. In vitro systems derived from organs and tissues from animals and humans, and various types of cultured recombinant cell lines, are successfully used for screening for the toxic potential of single compounds.

However, their ability to screen for the potential toxicity of whole foods or extracts thereof has been insufficiently explored up till now, and has been examined by us within the framework of a tiered safety evaluation of GM plants. The validity of using in vitro test assays for comparative testing of whole food products, food ingredients and extracts thereof depends on the sensitivity and selectivity of the toxic response induced upon exposure. However, since these models are not yet validated, they should be considered as early warning or alert-systems to identify potential changes in toxicity of the engineered product only.

ASSESSING TRANSCRIPTION FACTORS

There is a demand for in vitro models that not only determine the (geno-)toxic potency of plant components or complex mixtures, but also provide information on the mechanism by which that compound exerts its toxic effect. As an in vitro profiling system the eukaryotic stress gene assay (*i.e.* CAT-Tox(L)) was assessed. This utilizes human HepG2 cells stably transfected with chloramphenicol acetyltransferase (CAT) reporter constructs. The aim of the CAT-Tox(L) assay is to collect and quantify molecular responses to cellular stress and toxicity that have a transcriptional component in their regulation. The assay employs an ELISA method to measure the transcriptional activity of the promoter or response element constructs. The advantage is that many of the stress responses occur before any measurable cytotoxicity, thus allowing monitoring of stress pathways at sublethal levels.

Since 14 unique transcription responses can be measured simultaneously, the CAT-Tox(L) provides comprehensive cellular stress profiles. The assay generates specific stress gene induction profiles for a variety of stressful and toxic plant components such as genistein, diadzein and sinigrin.

Aqueous extracts of red-ripe, non-and antisense RNA exogalactanase tomato fruit dosed up to 0.05 g per assay did not cause any cytotoxicity to the HepG2 transfectants nor were there molecular responses to cellular stress or toxicity. Extracts of non-modified and modified green tomato fruit showed cytotoxicity and induced the construct containing the xenobiotic response element (XRE).

This gene-induction profile was not comparable to that of, for example, the reference molecules dioxin, 3-methylcholanthrene, benzo[a]pyrene or e -napthoflavone. These polycyclic aromatic hydrocarbons were characterized with a co-induction of cytochrome P450 1A1 and glutathione-S-transferase Ya subunit- CAT fusion constructs. There was substantial evidence that the systemic fungicides fenarimol and bupirimate did not induce any of the genes in this assay. There was no promotor in the assay that could be linked to universal cellular stress.

A-tomatine caused severe cytotoxicity at high concentrations and yet did not induce any of the genes (*i.e.* causing membrane damage). However, it was observed that d-tomatine diluted in a matrix of red-ripe tomato tissue (final extract concentration 5% w/v) elicited an increased cytotoxicity concomitant with a dose-dependent response profile involving XHF (collagenase involved in inflammatory reactions sensitive to mitogens and cytokine IL-1). So far, the approach of the CAT-Tox(L) model shows potential but there is a need for validation based on molecular mechanisms of plant constituents.

Although the induction of these stress genes represents cellular perturbations, the mechanistic validations need to be extended before any conclusions can be drawn on the value of these findings. Certainly, the development of a database to explore relationships between a wide variety of compounds stressful to plants and meaningful stress gene profiles may serve as molecular fingerprints unique for a particular component.

GENOTOXICITY ENDPOINTS

The mucosa of the gastrointestinal tract is clearly a primary, potential site of action for the effects of novel foods. Therefore effort has been concentrated on the development of a battery of intestinal cell systems (*e.g.* IEC-6, IEC-18, Caco-2 and INT407 cell lines) with different endpoints for initial cyto- and genotoxicity testing.

The choice of a suitable method of sample preparation was considered of utmost importance in assaying vegetables.

Freeze-dried tomato fruit was extracted (10% suspensions, w/v) in water, 0.9% saline or chloroform:methanol (2:1, v/v) immediately before incorporation into the culture medium. It was encouraging that aqueous and chloroform/ methanol extracts of, for example, red-ripe tomatoes up to a maximum of 10% suspensions (w/v), had little or no general toxicity once the pH (> 6.0) and osmotic pressure had been taken into account.

No toxicity to the intestinal cells Caco-2 and IEC-18 under short-term conditions was detected in red-ripe, non- and antisense RNA exogalactanase tomato using alpha-glutathione S-transferase (d-GST) lactate dehydroyrenase (LDH) leakage, reduction of 3-(4,5-dimethylthiazol-2-yl)-2,5 diphenyltetrazolium bromide (MTT conversion), neutral red (NR) uptake and total cellular protein as in vitro endpoints.

On the other hand, green tomato fruit displayed a pronounced toxic and cell-damaging effect. No statistical significant differences were found between transformants and isogenic control lines in terms of their activity in the assays. Treatment of tomato tissue with pepsin (ratio pepsin to tomato protein 1:250, w/w) did not result in an increase of or loss of inherent cytotoxicity. Supplementary investigations with naturally occurring plant components, for example, quercetin, e -carotene, d-lycopene, genistein showed that only steroid glycoalkaloids, such as d-tomatine, exhibited a rapid toxic effect, whereas d-tomatidine was ineffective. It was noted that the in vitro models did show general matrix-related toxic effects.

For example, d-tomatine diluted in a matrix of extracted red-ripe tomato tissue (final concentration of 5-10% w/v) showed an increase of dose-response-dependent cytotoxicity in human Caco-2 cells (*e.g.* IC50 7.9 ± 2.1 mg/L versus IC50 33.9 ± 5.3 mg/L of d-tomatine only). Using the COMET assay, suspensions of freeze-dried tomato fruit up to a maximum of 10% (w/v) did not exhibit DNA-damaging effects in a range of gastrointestinal cells derived from rats (IEC-6 and IEC-18) and humans (Caco-2, INT407). Genetic modification did not result in increased genotoxicity, but fruit extracts were able to suppress the DNA-damaging effects of the known genotoxins H2 O2 and N-methyl-N2 - nitro-N-nitrosoguanidine (MNNG). Although these endpoints are important toxicological measures they do not indicate the molecular mechanisms of toxic damage. So far, pronounced adverse effects could only be observed with glycoalkaloids. The screening for subtle differences in toxic responses of other plant constituents appeared to be limited. However, therefore, the use of critical endpoints for such studies other than cytotoxicity should be clearly defined.

POTENTIAL APPLICATION OF DNA MICROARRAY TECHNOLOGY

A second potential application of DNA microarray technology in safety evaluations concerns screening for biological effects of (fractions of) plant compounds that have been found by other profiling techniques to be present or modified in the genetically modified crop plant when compared to the parental line and other references. At RIKILT-DLO we are testing the feasibility of this strategy by analysing defined food compounds for possible effects on human intestinal gene expression. To this end, human intestinal cell lines are exposed for different time periods to various amounts of crop plant components and the mRNA isolated. The labelled mRNA is hybridized to DNA microarrays containing unknown intestine-specific genes, as well as already identified control genes known to be intestine specific or to be functional indicators (for toxicological processes, apoptosis, etc.).

As a source for unknown, intestine-specific genes we use cDNA libraries from human intestine biopsies and human intestinal cell lines which have been enriched by a subtractive cloning approach. Having established the usefulness of this approach for the identification of biological functions of defined plant compounds, these cDNA microarrays will be exploited to screen, for example, GM crop plants for possible effects on human health. The concept of substantial equivalence is broadly accepted as a basis for the risk assessment of GM crops and derived novel foods.

Data requirements for establishing substantial equivalence should be based on identifying unintended metabolic perturbations in the engineered crops that have been caused by genetic modification. A generic approach cannot be envisaged, however. In all cases the collation of data should be directed by the type of genetic modification and related consequences. It is recommended that unintended effects should be identified on a case-by-case basis, since such effects are dependent on the genotypic and phenotypic parameters of the new gene products involved.

The application of innovations in molecular genetics research will help to define the conditions under which the new food products can be marketed. The present approach has generated a sound scientific basis for the evaluation of the potential of unintended effects in GM crops and derived food products. The results may assist regulators and legislators to test and, if necessary, to improve the applicable regulations.Moreover, the technologies described and the results obtained may serve as a general framework for the risk assessment of GM crop plants, and may contribute to a better understanding by the public of recombinant DNA techniques in plant breeding and their implications. It is concluded that a parallel approach focusing on a comparative analysis using informative profiles of molecules at various integration levels (*e.g.* mRNA, protein, metabolites) including a toxicological profiling of relevant plant matrices, offers good prospects for the identification of hazards related to unintended effects. The choice of comparators and of external parameters to support a claim that a GM crop plant presents no unintended effects compared with its traditional counterpart should be based on sound scientific judgement. It is recommended that data are generated on the unprocessed part of the plant, such as the tomato fruit or potato tuber, used for human and/or animal consumption.

Furthermore, data banks should be set up containing information on natural variations of essential plant or other product constituents (*i.e.* crop mean values). The chemical fingerprinting technique, using a combination of off-line liquid chromatography and proton-NMR imaging, appears to be a powerful screening method for the detection of secondary effects in the metabolome that may be applied on a routine basis.

The carbohydrate profiling technique is useful for the detection of unexpected changes at N-glycan levels; in particular it is applicable to the detection of post-translational modifications in newly expressed proteins as well as in the whole GM crop plant. DNA microarray technology is rapidly evolving and much effort is being put into improving spot density, reduced production time, and increased reproducibility and sensitivity. Improving the latter is critical because it makes it possible to use smaller amounts of starting material. It will depend on technical aspects such as the quality of the scanners and spotting machines, the development of fluorescent dyes with improved characteristics (*e.g.* narrow excitation and emission peaks; high level of photon emission; resistance to photo-bleaching), supports with reduced background and more target sequence binding capacity.

DNA microarrays generate a huge amount of complex hybridization data and major challenges are to develop more advanced computer software to help to find statistically significant correlations within, and between different experiments, and to link the data to sequence information and submitted expression profiles available in public databases. Application of these technologies in combination with other measurements of, for instance, the performance and quality may limit or even replace animal feeding studies aimed at the detection of unintended effects.Together, these techniques, known as functional genomics, will enable us to answer questions about what happens to expression and protein composition when a plant cell changes metabolic state, for example, due to genetic modification or environmental factors. It is foreseen that this holistic view on the cellular machinery will be achieved soon, as the coincidence of genome sequencing with improvements in the analysis of expressed proteins is of growing importance and continues to be successful.

Eventually, microarrays for genomic studies will probably also be used to evaluate the numerous constituents of GM crop plants. Some constituents might not cause obvious changes in cellular behaviour or morphologic appearance but could cause subtle metabolic alterations that would show up when the mRNA content was interrogated by an array. It is likely that microarrays might be used to evaluate the life cycle of a plant much more precisely and to understand the complex metabolic control systems of plants. The ultimate test for GM crop plants will be acceptance by the public in large.

It is therefore extremely important that industry, the scientific community and the regulatory authorities are able to respond to queries from the public by providing transparent information on the criteria for safety assessment. Improved safety assessment at the molecular level should refine and complement strategies by anticipating the potential risks and sources of unintended effects in GM plants.

GENETICALLY MODIFIED FOOD CROPS

Genetic modification, otherwise referred to as recombinant DNA (rDNA) technology or gene-splicing, has proven to be a more precise, predictable and better understood method for the manipulation of genetic material than previously attained through conventional plant breeding. To date, agricultural applications of the technology have involved the insertion of genes for desirable agronomic traits (*e.g.* herbicide tolerance, insect resistance) into a variety of crop plants, and from a variety of biological sources.

Examples include soybeans modified with gene sequences from a Streptomyces species encoding enzymes that confer herbicide tolerance, and corn plants modified to express the insecticidal protein of an indigenous soil microorganism, Bacillus thuringiensis (Bt). A growing body of evidence suggests that the technology may be used to make enhancements to not only the agronomic properties, but the food, nutritional, industrial and medicinal attributes of genetically modified (GM) crops.

Regulatory supervision of rDNA technology and its products has been in place for a longer period of time in the United States than in most other parts of the world. The methods and approaches established to evaluate the safety of products developed using rDNA technology continue to evolve in response to the increasing availability of new scientific information.

As our understanding of the potential applications of the technology is broadened, the safety of products developed using rDNA technology and the potential effects of introduced gene sequences on human health or the environment will be more closely scrutinized. In fact, much of the knowledge acquired during the commercialization of the products of rDNA technology in agriculture is now finding application in evaluating the safety of products developed through more conventional means. The objective of this chapter is to provide the reader with an overview of the significant events leading up to the present, science-based, regulatory framework that exists for the safety evaluation of GM food crops within the United States. An attempt has been made to discuss concerns over the sufficiency of existing regulations, as well as to highlight recent initiatives taken by federal regulatory agencies to address them.

Through better communication of how the regulatory process functions within the United States, it is anticipated that current and future applications of rDNA technology in agriculture will be met with a greater level of understanding and acceptance.

HISTORICAL PERSPECTIVE

rDNA technology was first developed in the 1970s. The initial response of the scientific community, including members of the National Academy of Science (NAS), to the prospects of rDNA technology, was to postpone any further research involving the technology until the potential risks to human health and the environment could be evaluated. Researchers attending the International Conference on Recombinant DNA Molecules in 1975, otherwise known as the Asilomar Conference, tried to establish a scientific consensus on how best to self-regulate emerging applications of the technology. The conditions and restrictions that were proposed at this conference have formed the basis by which federal guidelines and policies for rDNA technology research were drafted within the United States.

National Institutes of Health (NIH)

The National Institutes of Health (NIH) was the first federal regulatory agency to publish their interests in evaluating the safety of rDNA technology in 1976, in the form of guidelines for the conduct of research. Because of the uncertainties that existed at the time, all research into the potential applications of rDNA technology was limited to the confines of federally funded labouratories under NIH control. After continued research, and a more careful assessment and monitoring of the risks, a set of less restrictive guidelines was published in 1978.

However, the environmental release of organisms developed using rDNA technology outside the confines of controlled labouratory conditions was prohibited unless otherwise approved by the NIH director. In the early 1980s, the NIH established an rDNA Advisory Committee (RAC) to review all data and experience gained with applications of the technology under its control. Based on recommendations of the RAC, a more relaxed set of research guidelines was published by the NIH in 1983.

The NIH approved the first environmental release of an organism developed using rDNA technology (ice-minus strain of Pseudomonas) in 1983. In response, they were criticized for failing to prepare a statement or assessment of the environmental impact of their regulatory decision as required under the National Environmental Policy Act (NEPA). Once the legal controversy had subsided, all responsibility that the NIH had for regulating the environmental introduction of GM organisms was relinquished. Nevertheless, NIH guidelines continue to be referenced in assessing the safety of rDNA research performed within industry, federal and other state labouratories. However, it was unclear which federal regulatory agencies would be responsible for ensuring the safety of the products developed using rDNA technology.

FOOD AND DRUG ADMINISTRATION (US FDA)

Authority

The US FDA is responsible for ensuring the safety and wholesomeness of all food and food components, including the products of rDNA technology, under the FFDCA. The US FDA has the authority for the immediate removal of any product from the market that poses potential risk to public health or that is being sold without all necessary regulatory approvals. As a result, a legal burden is placed on developers and food manufacturers to ensure the commodities utilized and foods available to consumers are safe and in compliance with all legal requirements of the FFDCA.

In order to understand the regulatory approach followed by the US FDA in the safety evaluation of GM crops, it is useful to consider food and food safety from a historical context. People had been consuming foods derived from agricultural crops for many years prior to the existence of any food laws or regulations within the United States. Based on this experience, agricultural crops have been accepted as being safe for consumption as food, without additional testing to demonstrate their safety. As long as the new crop variety has exhibited similar agronomic properties, and an appropriate taste and appearance, it has been considered safe to consume. As a result, most foods consumed today, in particular whole foods (*i.e.* fruits, grains and vegetables) and conventional foods, have not been subject to any kind of premarket review or approval by the US FDA. Nevertheless, food scientists have a good understanding that many of the commonly consumed agricultural crops contain natural toxicants (*e.g.* tomatine in tomatoes, solanine in potatoes, cucurbiticin in cucumber, psoralens in celery, etc.). As a result, new plant varieties may be subject to routine chemical analyses to ensure that none of these substances is present at potentially harmful levels. This type of general approach has been used in assessing the safety of thousands of new plant varieties that have been developed over a number of decades of crop breeding without compromising the safety of whole foods.

Role

Consistent with recommendations within the Coordinated Framework, the US FDA considered existing provisions of the FFDCA to be sufficient for the regulation of foods and food components developed using rDNA technology. It was concluded that the scientific and regulatory issues posed by the products of rDNA technology were not significantly different from those posed by conventional products.

As a result, GM foods and food components have been subject to the same standards of safety as already exist for the regulation of other foods and food components under the FFDCA. In order to better communicate interpretations of existing provisions of the FFDCA as they relate to the safety evaluation of foods derived from new plant varieties, including the products of rDNA technology, the US FDA released a policy statement in 1992 entitled 'Statement of policy: foods derived from new plant varieties'.

The US FDA has considered the use of genetic modification (*i.e.* rDNA technology) in the development of new plant varieties to represent a continuum of conventional plant breeding practices (*e.g.* mutagenesis, hybridization, protoplast fusion, etc.), and as a result, the safety evaluation of all new plant varieties, not just those developed using rDNA technology, have been evaluated based on an objective analysis of the characteristics of a food or its components, and not on its method of production.

OFFICE OF SCIENCE AND TECHNOLOGY POLICY (OSTP)

In response to a need for clarification, the Office of Science and Technology Policy (OSTP) began work on the development of a policy to establish a federal regulatory framework for evaluating the safety of products developed using rDNA technology. Following an opportunity for public comment, OSTP published a final version of the 'Coordinated Framework for Regulation of Biotechnology' (Coordinated Framework) in 1986.

The policy provided the basis by which federal regulatory agencies got involved in evaluating the safety of products at later stages of commercial development at that time. According to the Coordinated Framework, the products of rDNA technology should be regulated on the basis of the unique characteristics and features that they exhibit, not their method of production. The products of rDNA technology were considered to pose risks to human health and the environment similar to those posed by conventional products already regulated within the United States. As a result, no new federal regulatory agencies or regulations were required. The Coordinated Framework did not, however, rule out the possibility of the development of new guidelines, procedures, criteria or even regulations to supplement or alter the scope of existing statutes for the products of rDNA technology. The Coordinated Framework identified three federal regulatory agencies within the United States: the US Food and Drug Administration (US FDA), the US Department of Agriculture (USDA) and the US Environmental Protection Agency (US EPA), as having primary responsibilities for evaluating the products of rDNA technology under development at that time. In 1992, the OSTP released another document entitled, 'Exercise of Federal Oversight within the Scope of Statutory Authority: Planned Introductions of Biotechnology Products Into the Environment', outlining the proper basis by which federal regulatory agencies were expected to exercise their regulatory authority. As with conventional products, dependent upon the intended use and function, more than one federal regulatory agency may share an interest in evaluating the safety of a product developed using rDNA technology.

If more than one federal regulatory agency has an interest, lead agencies are identified as being responsible for coordinating activities to limit any potential duplication of efforts. Although federal regulatory agencies worked independent of one another, it was realised that close working relationships would need to be established in order to evaluate effectively the safety of products developed using rDNA technology.

Recently, the OSTP teamed up with the White House Council on Environmental Quality (CEQ) to perform a six-month inter-agency evaluation of the federal regulatory agency responsibilities in evaluating the environmental safety of products developed using rDNA technology. A case-study approach for a variety of different classes of products developed using rDNA technology was used to evaluate the level of federal regulatory agency involvement, to identify strengths, weaknesses and areas of potential improvement. The review concluded that none of the previously approved products of rDNA technology has had any significant negative impact on the environment.

Although all the case studies were published, OSTP/CEQ failed to reach a consensus on issues relating to the relevant strengths and weaknesses of the existing regulatory structure within the time allotted for the completion of its review. A review of the case studies published provides a comprehensive interpretation of the responsibilities of each federal regulatory agency in ensuring the safety of the products developed using rDNA technology.

NATIONAL ACADEMY OF SCIENCES (NAS)

The National Academy of Sciences (NAS), and its operating arm, the National Research Council (NRC), have served as a primary source of scientific, technological, human health and environmental policy advice during the development of regulatory approaches for the safety evaluation of products developed using rDNA technology within the United States. In 1987, the NRC published a report concerning the potential human health and environmental hazards associated with the commercial introduction of GM organisms, entitled Introduction of Recombinant DNA-Engineered Organisms into the Environment: Key Issues. The risks associated with the introduction of GM organisms were considered to be essentially the same in kind as those associated with unmodified organisms. In other words, rDNA technology did not appear to introduce any unique risks as compared to the products that had been developed using more conventional methods of genetic modification. In reaching these conclusions, the NRC performed an evaluation of the similarities and differences in the properties exhibited by products developed using a variety of different techniques. To this day, the conclusions of this report continue to be referenced by the regulators and developers of GM crop varieties worldwide. A subsequent NRC report, entitled Field Testing Genetically Modified Organisms: Framework for Decisions, also reached similar conclusions, but provided additional guidance as to how regulatory decisions concerning the introduction of GM organisms should be made. The NRC recommended that regulatory decisions concerning the introduction of GM organisms should be made on a case-by-case basis.

Consistent with the Coordinated Framework, the NRC did not consider the nature of the process used for the genetic modification of an organism to be a useful criterion for determining whether a product requires less or more regulatory oversight. As a result, no valid reason existed to regulate organisms genetically modified via modern techniques (*e.g.* rDNA technology) any differently from organisms genetically modified via more conventional means.

Similar conclusions have been published in the reports of international standards-setting organizations. In retrospect, within both reports, the NRC acknowledged that modern and conventional methods of genetic modification are not without risks to human health or the environment, and as a result, neither could be considered inherently more risky. As a result, regulatory decisions concerning the safety of the products of rDNA technology need only to take into consideration the specific characteristics exhibited by a particular GM organism and the environment in which it is to be introduced, and not the method by which it has been produced.

The NRC has further articulated their conclusions into what is now commonly referred to as the 'Concept of Familiarity'. Although familiarity with the characteristics of a particular organism or the environment to which it will be introduced would not necessarily mean it was safe, it can be expected to provide a sufficient amount of information to allow for a judgement to be made of the risks.

For example, familiarity with a new GM plant variety could be established based on comparisons between characteristics of the parent line or other crop species exhibiting similar traits, as well as through the results of actual field tests involving the GM plant. These principles were further elabourated upon by the Organisation for Economic Co-operation and Development (OECD), and as a result, have been referenced in the development of regulatory policies for evaluating the safety of GM crops on a global basis.

More recently, a committee established by the NRC published a report entitled Genetically Modified Pest-Protected Plants: Science and Regulation, based on a review of all scientific and regulatory data collected during the regulatory approval process for GM crops within the United States. The primary objective was to assess independently the effectiveness of existing and proposed regulations for the safety evaluation of GM crops expressing plant pesticides (*i.e.* plant-incorporated protectants).

No new evidence was identified to suggest plants expressing plant pesticides posed any greater risk to human health or the environment as a result of their genetic modification. In fact, the NRC concluded, 'with careful planning and appropriate regulatory oversight, commercial cultivation of GM plants is not expected to pose higher risks and may pose less risk than other commonly used chemical and biological pest-management techniques'.

However, the NRC report included requests for federal regulatory agencies to further strengthen the current regulatory approval process through better coordination and communication between agencies, on-going investment in the research and monitoring of potential human health (*e.g.* allergenicity) and environmental impacts (*e.g.* insect resistance), and by providing greater access to information evaluated in support of regulatory decisions.

US FEDERAL REGULATIONS FOR AGRICULTURAL BIOTECHNOLOGY

Regulatory systems for the products of agricultural biotechnology have been in existence since the mid- to late 1980s within the United States. The regulatory approach to the safety evaluation of plants developed using rDNA technology has evolved in the best interests of research scientists, industry and the general public. The agricultural products of rDNA technology, such as GM foods and crops, may require approvals from up to three regulatory agencies; the US FDA, the USDA and the US EPA, depending upon the characteristics exhibited by the GM plant, its proposed use and introduced traits. The same standards of safety are applied to all products regardless of the technology used in their development. The US FDA is responsible for ensuring the human safety of all new foods and food components, including products developed using rDNA technology, under the Federal Food, Drug, and Cosmetic Act (FFDCA).

The USDA evaluates the potential of a GM plant to become a plant pest following its environmental introduction under the Federal Plant Pest Act (FPPA). The US EPA evaluates pesticides, including plant systems modified to express pesticides (*e.g.* insect-protected or virus resistance), under the Federal Insecticide, Fungicide, and Rodenticide Act (FIFRA).As a result, the expression of an insecticidal protein in a food crop would undergo review by the USDA, US EPA and US FDA; a GM food crop exhibiting a modified oil content would be evaluated by the USDA and US FDA; and a non-food horticultural plant developed using rDNA technology for any other purpose (*e.g.* flower colour) would be subject to review by the USDA alone.

In most instances, obtaining all necessary approvals for the commercialization of an agricultural crop developed using rDNA technology takes a decade or more. However, the exact amount of time required will depend on the need to confirm performance, to evaluate characteristics of the food, environmental effects, and to produce the required amount of seed before the product can be distributed and commercially grown by farmers.

Up to five years of field trials (5-10 generations of plants) are required for the developer of a new plant variety to collect sufficient data to meet the reporting requirements of the USDA. An additional five months to two years may be required for the US FDA, USDA and/or US EPA to complete all necessary product consultations, reviews and approvals. Approval for the first commercial planting of a GM food crop was not issued until 1995. Since 1995, more than 40 new agricultural crops developed using rDNA technology have received approval for commercial planting within the United States.

In 1999, approximately 72% of the total 39.9 million hectares (more than 98 million acres) of GM crops grown worldwide were planted in the United States. Herbicide-tolerant soybeans (54%), Bt corn (19%) and herbicide tolerant canola (9%) accounted for approximately 82% of the GM plants cultivated. With an increasing number of agricultural biotechnology products reaching later stages of commercial development, it is anticipated that the overall area planted with GM crops will continue to rise.

The regulatory approach to evaluating the human health and environmental safety of GM crops within the United States is best described as a science-based, case-by-case assessment of hazards and risks. This approach has provided the flexibility required to reduce the regulatory burden placed on products that have been determined to be of low risk or concern.

All agencies involved in the regulation of plants developed using rDNA continue to implement and develop policies based on recommendations made within the Coordinated Framework. In the future, the US FDA, USDA and the US EPA will be dedicating additional resources towards communicating how GM food and food components are regulated within the United States, and how these regulations function to be protective of both human health and the environment.

METHODS OF TRANSFORMATION IN PRODUCT CHARACTERIZATION

Product characterization allows for consideration of any potential hazard(s) that might be associated with anticipated exposure to a plant-incorporated protectant. This process includes a review of the source of introduced genetic material, including all regulatory sequences, the modifications that have been made, methods of transformation, inheritance, stability and expression in the host plant.

For the introduced pesticidal substance, the anticipated modes of action, specificity and toxicity to both target and non-target organisms are evaluated. Target organisms may include weeds, insects and diseases (viral, bacterial, and fungal) affecting a crop.

The crop itself, as a volunteer weed, may also be of concern following crop rotations. Once a PIP and any potential for exposure have been characterized, the data required in support of an environmental and human health risk assessment can be defined.

ECOLOGICAL EFFECTS

Environmental fate and exposure are important considerations when evaluating the potential impact of PIPS on the environment or on non-target organisms (*e.g.* wildlife, beneficial insects). Depending on the crop and introduced pesticide substance, data on the level of expression and rate of degradation may be required, first for the tissues of the modified plant, and secondly, the fate in soil, water, or any other environmental compartment.

Such data allow the potential for exposure to the pesticide substance to be evaluated for non-target organisms via feeding on, or through decomposition of, the modified plants. Studies are arranged in tiers, starting with Tier I, consisting of acute toxicology and basic characterization studies that help predict environmental fate. Depending on the results obtained in Tier I, and the level of concern, additional studies may be requested from higher tiers. It is expected that most PIPS will only undergo Tier I testing, since negative results are expected at relatively high exposure levels.

The 'maximum hazard approach' when used in Tier I provides a high level of assurance that adverse effects would not be expected to occur under actual conditions of use within the field. If the results of Tier I reveal a potential hazard at doses within the range of expected concentrations in the environment, additional studies would be requested in higher tiers to better characterize the potential for any kind of effect. Data requirements may be satisfied by generating test data or by requests for data waivers with credible justification.

The type and extent of dietary studies required will depend on properties of the crop and on the introduced pesticide substance. Dietary studies typically involve feeding plant tissues (*e.g.* grain, pollen) that express the PIP or that have been spiked with the pesticidal substance. However, when determined necessary, the pure pesticidal substance may be administered to non-target species at doses of up to 10 to 100 times the expected level of exposure in the field. Dependent upon the potential for exposure, additional studies may also be requested involving any number of other non-target organisms (*e.g.* honeybees, green lacewing, ladybird beetles, parasitic wasps, earthworms, etc.). When toxicity is observed, the potential for exposure becomes an important consideration when ascertaining whether an adverse effect might be expected under actual conditions of use in the field. The requirement for provision of a resistance management plan is not routine practise for PIPS, but depends on an evaluation of the potential for resistance to develop and the threat to existing or functionally equivalent pest management practises. To date, resistance management plans have only been placed on Bt PIPS. If considered appropriate, the EPA may request that a resistance management plan be submitted as a part of a condition for registration. A resistance management plan may require a commitment towards the generation of additional research data, to implement structured refuges, to perform annual resistance monitoring, to initiate remedial action plans, and to educate growers. The EPA has worked closely with academia, other federal agencies, public interest groups, industry, and growers, to establish resistance management plans based on the most current science.

HUMAN HEALTH EFFECTS

For agricultural food crops, special consideration is given to evaluating the safety of the pesticidal substance introduced as a component of consumed foods. Rodents (often mice, to conserve test material) are orally administered the pesticide substance at doses of up to 100 000 times the levels of expected human consumption as a part of the diet.

Although the modified plant tissues would be considered the most appropriate vehicle for determining the acute toxicity of a PIP, it is not typically feasible, since only small amounts of the pesticidal substance are present in plant materials.

Therefore, an alternative source (*e.g.* microbial) is often used to generate a sufficient amount of identical pesticide substance for testing purposes. In this case, the equivalency of the test substance obtained from an alternative source to the PIP (*e.g.* composition, identity, etc.) being expressed within the plant is an important consideration that must be well documented.

For pesticide substances that are proteins, the principal concern relates to the potential for the introduction of a new allergen or toxin into the food supply. Since most allergenic food proteins are stable to digestion, in vitro digestibility studies are used to predict how long a pesticidal substance may persist in digestive fluids (*e.g.* gastric, intestinal) following ingestion. In addition, heat stability studies are performed to provide an indication of the fate of the

PIP during food processing. Commonly, assays of amino acid sequence homology are performed to determine similarity to substances that are known to pose unique safety concerns (*e.g.* toxins, allergens).

The EPA uses a 'decision tree' approach similar to that followed by the FDA in assessing the potential allergenicity of a new food protein. In many instances, US EPA and US FDA collabourate with each other when evaluating issues of safety for new pesticidal proteins being expressed in GM food crops.

Residue Tolerances for Plant-incorporated Protectants

Analogous to the consideration of tolerances for residues of conventional pesticides, the US EPA reviews data on the human, animal and environmental safety of PIPS to determine whether residue tolerances should be established for the amounts expected to be present in foods derived from the GM plant. Under Section 408 of the FFDCA, if the GM plant expressing a PIP is a food crop, the US EPA is obligated to establish the 'safe level' of pesticide residue allowed, or tolerance level.The Food Quality Protection Act (FQPA) served to amend the FFDCA and FIFRA by outlining the process by which the safety of pesticide residues on raw or processed foods can be established. Under this process, in order for a pesticide residue to be considered safe, a 'reasonable certainty of no harm' must exist assuming aggregate exposure of the public, taking into consideration sensitive children within the population.

To date, all pesticidal proteins that have been registered as PIPS have been determined to be non-toxic, and as a result, have been considered exempt from the requirement to establish a tolerance under the FFDCA. Consistent with categorical exemptions as originally proposed for plant-pesticides under FIFRA, the US EPA also proposed that tolerances might not be required for pesticidal substances derived from sexually compatible plants, regardless of the process used to introduce the PIP, provided that the genetic material encoding the pesticidal substance was from a plant commonly used as food, and that the presence of the pesticidal substance in a host plant did not result in any new or significantly different dietary exposures.

Final rules include exemptions from tolerances for the residues of PIPS, and their encoding nucleic acids, but only for PIPS derived through the conventional breeding of sexually compatible plants. US EPA is requesting further comment on the possible exemption from the requirement of a tolerance for:

- PIPS derived through rDNA technology;
- PIPS that act by affecting plant defences;
- PIPS developed based on viral coat proteins.

Future Considerations

The US EPA anticipates that the requirements for the registration of PIPS will continue to evolve as the science and policies relating to biotechnology continue to mature. In light of on-going criticisms that the plant-pesticide rule has received since its publication in 1994, a number of stakeholder workshops have been hosted by the US EPA. In March 1999, the House Agricultural Subcommittee hosted a congressional hearing to address a number of these concerns.

One of the primary questions addressed was whether the definition of a pesticide under FIFRA provides the US EPA with the authority to regulate pest-protected plants that have been developed using rDNA technology. Another concern was that the proposed rule was too broad in scope because all plants have natural defence mechanisms, and as such, would be classified as 'plant-pesticides'. In addition, an exemption from the plant-pesticide rule for those plants not developed using rDNA technology would serve to regulate the process used in the development of the GM plant, rather than on the pesticidal substance being produced within the plant. Further criticism related to the use of the term 'plant-pesticide' in relation to the products of the genes that had been introduced into GM crops.

As a part of its final rule, the US EPA recommended that plant-pesticides be renamed PIPS, an alternative name which distinguishes them from the other types of pesticide products requiring registration with the US EPA. A new FIFRA section has been added to the Code of Federal Regulations specific to the regulation of PIPS. In doing so, the US EPA has acknowledged that the new final rule would be making a distinction between the products of conventional breeding and rDNA technology, *i.e.* regulating on the basis of process not product.One reason given for this divergence was that the use of this criterion would be expected to provide the public with increased confidence that an appropriate level of regulatory oversight was in place for assessing the risks of rDNA technology, which is a societal not scientific issue. SAPs are charged with providing the Office of Pesticide Programmes (OPP) of the US EPA with advice and guidance on human health and environmental issues relating to proposed regulatory decisions for pesticides, including PIPS, based on the latest scientific data. For example, the SAP for Bt Plant Pesticides met three times in the year 2000 to consider issues related to the safety of plants that had been genetically modified to express insect-specific Bt toxins. The US EPA released a decision to temporarily extend registrations for all currently registered pest protected plants throughout the 2001 growing season.

Additional provisions, such as the recently strengthened resistance management requirements for plant-pesticides, along with the original registration conditions, were considered to provide adequate protection during the extended time period. In addition, the reassessment process for these products has been designed to assure maximum transparency and opportunity for public comment in the regulatory decision-making process.

PROTECTION OF DOMESTIC AGRICULTURAL RESOURCES

The Animal and Plant Health Inspection Service (APHIS) of the USDA is responsible for the protection of domestic agricultural resources and the environment from the threats posed by pests and diseases. The Federal Plant Pest Act (FPPA) has been the primary statute adapted by APHIS in regulating the products of agricultural biotechnology. Under the FPPA, APHIS has the authority to restrict the import of plant pests, to introduce measures (*e.g.* quarantines) to prevent their movement, or to destroy plant pests cultivated in violation of the FPPA. The requirements under the FPPA are equally applicable to scientists within company, academic, research, and private organizations.

Definition of Plant Pests

A plant pest is any agent, substance or organism, or parts thereof, that may cause injury, disease or damage, either to plants or to the environment in which it is introduced. APHIS maintains a list of organisms considered to be plant pests, and therefore, subject to regulation under the FPPA. If an organism is not on the list, it may still be subject to regulation as a plant pest if there is reason to believe it is or will act as a plant pest. Organisms considered plant pests, or that exhibit the potential to be plant pests, are treated as 'regulated articles' under the FPPA. The regulations extend to the introduction of GM organisms meeting the definition of a plant pest or for which there is reason to believe they are plant pests. APHIS considers the environmental release of a GM crop to be equivalent to the introduction of new organisms.

Therefore, until proven otherwise, a GM crop is considered a 'regulated article' under the FPPA. The use of the term 'plant pest' in reference to GM plants means the 'non-pest' nature of the crop has yet to be determined. To date, all field tests conducted have demonstrated that GM crops exhibit no more 'pest-like' traits than their conventional counterparts. All GM plants developed to date have involved the use of at least one designated 'plant pest' as either promoters or vectors, and as a result, have been subject to regulatory review under the FPPA.

APHIS takes the position that an entire plant can be designated as a regulated article, even if it was not developed using plant pests, if the plant is the product of genetic engineering, and the agency determines or has reason to believe it is a plant pest. Given this position, APHIS would likely challenge any attempt to introduce a GM plant into commerce unless the developer has first gone through the APHIS regulatory review process.

Requirements for Commercialization

A number of years of field testing are required to evaluate the agronomic and product quality characteristics exhibited by new plant varieties developed within a labouratory or greenhouse. APHIS regulations outline procedures for obtaining a permit or providing notification prior to the importation, interstate movement or release of a regulated article, which of course, includes field testing. Prior to an introduction, a proponent must provide notification to APHIS of its intentions or submit an application to obtain a permit. The notification and permit requirements have evolved as an extension of the long-standing programme for the regulation of plant pests within the United States. Since APHIS has issued permits or acknowledged the receipt of notifications for the field release of more than 6500 new crop varieties.

Permits

For APHIS approval, developers must demonstrate that a new plant variety poses no significant risk to other plants in the environment or is at least as safe as similar crop varieties. APHIS review and approval are required for the shipment of seed and for the conduct of field trials involving new crop varieties. APHIS issues site-specific permits for field tests or releases into the environment.

Prior to field testing, APHIS may seek further clarification of study objectives, specify how, when or where research may be conducted, establish the data to be generated, collected and reported, stipulate additional requirements for the monitoring or securing of test sites, and specify methods for the disposal of crop residues that remain at the end of a study.

The results of field trials are often used in making a determination as to whether a GM plant exhibits any effects on non-target species, or poses any unique plant pest problems. For field trials conducted under either notification or permit, developers of GM plant varieties are required to submit reports to APHIS that include: details regarding the methods of observation; the data collected; and interpretations concerning the effects on plants, non-target organisms, or the environment.In the event that a GM crop is likely to be the subject of a petition for a determination of non-regulated status, developers must provide a description of known and potential differences, and substantiate that the regulated article is not likely to pose a greater plant risk than the organism from which it was derived. An environmental assessment is a requirement of the NEPA, Council on Environmental Quality regulations, and USDA procedures for issuing permits under the FPPA. Under the permit process, developers of GM plant varieties must disclose information concerning the plant, test facilities, and control measures in place for its transport and field testing. Based on this information, the USDA performs an assessment of the potential environmental impact following a release. If the agency reaches a 'Finding of No Significant Impact' (FONSI), a permit may then be issued.

Notifications

Depending on the plant species and intended introduction, applying for a permit from APHIS can be a complicated and time-consuming process. In the spring of 1993, APHIS introduced a simplified notification process as an alternative to applying for a permit, which in most instances applies to the introduction of GM plants. To qualify for notification, a GM plant must be introduced in accordance with the eligibility criteria. Originally, six GM crops (corn, cotton, potatoes, soybean, tobacco and tomatoes) qualified for notification; however, the process was more recently expanded to include most other crop species not capable of becoming a noxious weed. The developers of new GM plant varieties are encouraged to contact APHIS when making a determination as to whether a GM plant would qualify for notification. If a particular GM plant does not qualify for notification, a more involved regular permit process must be followed.

For a GM plant to qualify for notification, it and its predecessor must not be a noxious weed, the introduced gene sequence(s) must be stably integrated within the host, not originate from a plant or animal pathogen, exhibit a known function that will contribute to plant disease, or encode for substances that could be potentially toxic or infective to non-target organisms, such as pharmaceuticals, or viral components other than those coat proteins already established as safe.Performance standards also exist for the conduct of field trials to ensure that an adequate level of containment is provided for all introductions. It is the responsibility of the developer to ensure that an adequate level of containment is provided for field trials involving GM plants under notification. The performance standards utilized will vary according to the biology of the plant and the nature of the introduction. In general, this includes specifying methods for the proper handling, shipment, field monitoring and disposal of plant material at the end of the study.

As a part of the notification process, an applicant must provide information about the plant, identify the source of any genes used; the method of genetic modification; and the size, date and location of any proposed field test or introduction. Notification is required at least 10 days prior to the interstate movement or 30 days in advance of the field testing or importation of a regulated article that qualifies for notification with APHIS.

Determination of Non-regulated Status

Once sufficient data have accumulated through labouratory and field trials performed under permit or notification, a developer may petition APHIS for a determination of non-regulated status. A DONRS allows a GM plant to be grown, tested or used for crop breeding without further regulatory oversight by APHIS. APHIS will grant a DONRS if sufficient evidence can be provided to demonstrate that no plant pest potential exists (*i.e.* that the GM plant is not expected to become a pest, poses no significant risk to the environment, or is as safe as conventional plant varieties). The developer is required to demonstrate a lack of change in disease or pest resistance status, an absence of any potential for contributing to development of a new plant pathogen or pest, as well as to address any questions relating to potential environmental consequences of the release. APHIS examines several parameters when making a determination as to whether a GM crop should be considered a plant pest. Reporting requirements include submitting a detailed rationale for development; overview of the biology of the crop (competitiveness, survivability, dormancy); molecular characterization; protein expression; morphologi-cal/phenotypic characteristics; outcrossing/geneflow; weediness; insect and disease susceptibility; and, recombination potential (for viral genes).

Other considerations include assessing whether a GM crop could itself become a weed. In making this determination, factors that are considered include: how the seeds are dispersed in the environment; whether the seeds are capable of surviving over the winter; the potential for the development of volunteer plants; and whether these plants could reproduce into viable offspring.Another important consideration is whether the seeds or plants are capable of surviving outside of a managed agricultural environment (*e.g.* watering, fertilizer, etc.), and how the introduced traits could influence the viability of the crop under marginal conditions. APHIS also takes into consideration potential adverse effects on wildlife, including birds, beneficial insects and mammals, following the introduction of a GM crop. Some of this information comes from field trials that are conducted in multiple locations for several years.By observing crops growing under actual conditions of use in the field, scientists can compare insect populations existing within a field planted with GM crops to those coexisting in fields planted with a non-modified variety of the same plant. Field trials are also useful in identifying any changes in the plant physiology, such as height, colour, leaf placement, time of flowering, etc., that result from modifications in the genome of the plant.

Monitoring of these changes during field trials allows APHIS to evaluate how these changes might benefit or adversely impact the behaviour of wildlife in contact with these crops. Knowing that wildlife, such as deer, often feed on agricultural crops, APHIS also takes into consideration the nutritional content of the GM crop. Comparisons of the essential nutrient content are often made between the GM plant and its conventional counterpart.Other interests may include determining the impact of the modification on the levels of adverse factors (*e.g.* natural toxicants or anti-nutrients) present in the plant. APHIS maintains a list of all crops that have received a DONRS. At any time, APHIS retains the authority to prohibit the commercial planting of a GM crop if it is determined that it is becoming a plant pest.

PETITION PROCESS

Before a GM plant can be freely transported and commercialized, the developer must first petition the APHIS for a DONRS. A petition for a DONRS requires the submission of extensive data on the introduced gene construct, effects on plant biology and effects on the ecosystem, including the spread of the gene to other crops or wild relatives. Developing all data necessary to support a petition for a DONRS may take months or years depending upon the scientific issues that need to be addressed. Based on the information provided within the petition, APHIS evaluates the potential impact (*e.g.* toxicological) that the introduced modifications will have on non-target organisms in the environment, including threatened or endangered species. A petition for a DONRS must provide a rationale for the development and introduction of the GM crop, starting with genotypic and phenotypic information on both the host and donor organism(s).

This includes a detailed description of the methods of transformation, identifying all gene sequences (*e.g.* promoters, leader sequences, introns, selectable markers, etc.), their function, and potential for plant risk following introduction.

A combination of different analytical methods may be used to demonstrate the stable integration of an introduced gene sequence within the genome, its inheritance and expression in host plants. To the extent possible, the gene(s) of interest (*e.g.* resistance, marker genes), levels of expression, and im pacts on levels of inherent plant toxicants in plant tissues, under experimental and actual growth conditions in the field, must be characterized. Field performance studies (*e.g.* leaf morphology, pollen viability, seed germination, viability, insect susceptibilities, disease resistance, yield, etc.) are used to identify any differences or similarities in phenotypic characteristics exhibited by the GM and non-modified crop.

The extent of field performance data required in support of a petition for non-regulated status depends not only on the nature of introduced gene sequences, but also on the physiology of the host plant. Less extensive field performance studies are required for plants that are: highly domesticated (*e.g.* corn); self-pollinating (*e.g.* soybean); male sterile; those with high seed germination rates (> 90%); or unlikely to influence their potential weediness or fitness (*e.g.* delayed ripening, oil seed content).

Whereas, GM crops exhibiting a tolerance or resistance to cold, salt, biotic (*e.g.* insects, pathogenic agents), or other abiotic (*e.g.* herbicide) stresses, are subject to more extensive field performance testing. APHIS considers several parameters in reaching a DONRS, including whether the GM plant can cross-pollinate with other plants in the wild, and if it can, the ecological consequences (*e.g.* insect resistance, herbicide tolerance) that might result.Extensive databases are maintained on species capable of cross-pollinating with GM crops (*e.g.* corn, soybean, cotton, canola, etc.), based on the results of years of breeding experiments, biological surveys and research conducted by agricultural and weed specialists within APHIS. In instances where out-crossing might be expected to occur, APHIS may stipulate the planting of refuge areas to limit the potential for adverse ecological effects following commercial planting.

AGENCY RESPONSE

APHIS will grant a determination of non-regulated status if the developer can sufficiently demonstrate a lack of plant risk. In making a determination that a GM plant will no longer be regulated, APHIS prepares two documents, an environmental assessment and a DONRS to satisfy regulatory requirements under NEPA and FPPA, respectively.

Both documents are developed based on the review of data submitted as a part of the petition for a DONRS. Once the DONRS has been granted by APHIS, a GM crop no longer requires notification or permit prior to its movement or release within the United States. A review for DONRS is generally completed within 10 months of submitting a formal application to the APHIS, allowing sufficient time for publication in the Federal Register, and opportunity for public comment. Notices of all petitions for a DONRS are published in the Federal Register, and the public is given 60 days to comment for or against the petition, after which APHIS has up to an additional 180 days to either approve or deny it.

APHIS provides access to a considerable amount of information relating to approved field trials and petitions received for GM crops. The petition for a DONRS must include scientific details regarding the genetics of the plant, the nature and origin of the genetic material used, the potential for indirect effects on other plants, and any other information that could be considered unfavourable to the petition.

Extensions of Non-regulated Status

In certain instances, the non-regulated status of a GM crop may be extended to include other varieties of the same crop species, provided that changes in introduced gene sequences are insignificant, and no new plant pest issues would be expected.

For an extension, developers must substantiate that the new regulated article poses no serious issues meriting review under separate petition, by providing a precise description of the genetic modifications to the regulated article and a detailed comparison to modifications made in the GM plant previously granted non-regulated status by APHIS.

Field trials must be performed under notification or permit to demonstrate the similarities in the phenotypic properties. In support of multiple extensions, product identity standards may be submitted describing the genotypic and phenotypic properties exhibited by new plant varieties considered extensions of an existing DONRS for a GM crop.

Future Considerations

The USDA recently appointed an Advisory Committee for Agricultural Biotechnology that will work in conjunction with the NAS (Standing Committee on Biotechnology Food and Fibre Production, and the Environment) on the completion of a critical review of the regulations and policies used by APHIS in the commercial approval of GM crops.

The committee comprises scientists, farmers and representatives of consumer groups and seed companies, and is anticipated to take two years to complete its assignment. In addition to APHIS, other functional areas within the USDA are beginning to get more actively involved with the regulation of agricultural biotechnology. For example, the Grain Inspection Packers and Stockyards Administration (GIPSA) has made a commitment to work with farmers and industry groups in the development and validation of reliable test methods and quality assurance programmes to differentiate GM from non-GM commodities. As a part of this commitment, GIPSA intends to publish a proposed rule on the standardization of test methods and approaches for the detection and identity preservation of grains. GIPSA is also intending to offer accreditation services for labouratories testing grains for the presence of GM content, and to evaluate commercial test kits to ensure they can be considered accurate and reliable. It is expected that an administrative fee will apply to the accreditation services being offered by GIPSA.

The USDA also has made continuous commitments to funding research that will assist federal regulatory agencies in making science-based decisions concerning the safe introduction of GM organisms into the environment. Active areas of research include assessing the potential risks associated with the environmental introduction of GM plants (*e.g.* gene flow), the cumulative effects of large-scale commercial plantings, the potential interactions between GM and non-GM crop species, programmed resistance, and the development of statistical methodology and quantitative approaches to measuring the risks associated with the field testing of GM plants, etc.

Environmental Protection Agency

The EPA has authority over the registration of chemical and biological pesticides under the Federal Insecticide, Fungicide and Rodenticide Act (FIFRA). A pesticide is any product 'intended for preventing, destroying, repelling, or mitigating any pest'. All pesticides require registration by the US EPA prior to their distribution. As a part of the registration process, an applicant is required to submit information and data in support of the safety of a pesticide in its intended use.

In addition, US EPA is responsible for establishing acceptable tolerance levels for registered pesticides on or in raw agricultural food commodities under the FFDCA. At any point, US EPA has authority to amend or revoke a registration or residue tolerance, if an adverse effect is observed, or if the risks associated with the use of the pesticide are determined to be unacceptable.

Role

Under the Coordinated Framework, the US EPA was identified as being responsible for the review, assessment and registration of the products of rDNA technology that act as pesticides. Leading up to this publication, the US EPA had acquired considerable expertise in the registration of biological pesticides (*e.g.* microbial).

The US EPA was first involved in discussions regarding the potential risks of pesticides developed using rDNA technology in the early 1980s, during a series of public meetings involving the FIFRA Scientific Advisory Panel (SAP) and the Biotechnology Scientific Advisory Committee (BSAC). It was recognized that the potential risks posed by plants modified to express pesticidal components through rDNA technology were unique and warranted additional consideration relative to conventional pesticides.

For example, the potential for introduced genetic material to out-cross from a modified plant to a sexually compatible wild or weedy relative through pollen spread was one of the unique, theoretical risks posed by plant-pesticides. As a result, the EPA proposed a plant-pesticide rule in 1994, outlining their interests in the regulation of products.

A package of three proposed final rules was released early in 2001, redefining regulations for plant-incorporated protectants (PIPS), previously referred to as plant-pesticides. Essential elements of the proposed rules for the regulation of GM plants expressing pesticidal traits have been followed since 1994. The final rules provide further clarification of how PIPS will be regulated under FIFRA and FFDCA, including specific exemptions and became effective in September

Definition of Plant-incorporated Protectants (PIPS) or plant-pesticides

US EPA's final rule clarifies the regulation of whole plants that contain PIPS. When plants are used intentionally for controlling pests in some way, such uses meet the definition of a pesticide under FIFRA. Nevertheless, the agency exempts the plants themselves from regulation, focussing instead on the substances they contain, previously referred to as plant-pesticides and now known as PIPS.A plant-incorporated protectant (PIP) includes the substance produced by the living plant, and all genetic material necessary for its production (including promoters, enhancers, etc.). Thus, genes and substances (*e.g.* enzymes) responsible for the conversion of natural plant constituents into pesticidal substances would be considered PIPS.

To date, PIPS registered by the US EPA have been proteins and the genes required to make these proteins within the plant. Crops involved include potatoes, cotton, field corn, sweet corn and popcorn. Clearly, Bt toxins represent the most predominant class of PIPS that have been registered by the US EPA. Major reasons for this include the success of microbial products containing Bt and their long history of use for nearly 40 years. It has been estimated that combined, Bt corn, potato and cotton were cultivated on approximately 10 million acres in 1997, 20 million in 1998, and 29 million in 1999. As a result, less chemical insecticide has been applied, and there have been significantly higher crop yields. In addition, due to reduced opportunity for opportunistic fungal infections of insect-damaged corn, lower levels of mycotoxins have been produced, reducing the toxicological risk to both humans.

Exemptions

US EPA has concluded that living plants with pesticidal activity do not require a high level of regulatory scrutiny by the agency and thus, are exempt from regulation under FIFRA. The agency commented that focussing on the substance produced within the plant, *i.e.* the PIP eliminates any need for the US EPA to register plants, thus conserving agency resources and reducing potential overlap with other agencies, notably the USDA. For example, although whole plants that are used as biological control agents themselves, such as chrysanthemums, are exempt, substances that are extracted from plants, such as the insecticidal material pyrethrum, are not excluded from regulation under FIFRA.

Further exemptions have been proposed, based on familiarity and presence of the pesticidal substances in the food supply. Only one of the additional proposed exemptions has been included in the final rule, and all others have been proposed for further public comment.

Within the final rule, EPA has exempted PIPS and encoding genetic material originating from plants sexually compatible through conventional breeding but not through rDNA technology. The rationale for this exemption is that such substances could be bred, either naturally or through conventional plant breeding, from close plant relatives that share genetic material from a common gene pool. On the other hand, genetic modifications that were never before possible can be made using rDNA technology, involving the genes from sexually incompatible plant species.

Therefore, the exemption of plant-pesticides developed using rDNA technology, involving genes from sexually compatible plants has not been finalized as a part of the supplementary proposal to the final rules for PIPS. Other exemptions for further comment have also been identified within the supplemental proposal. The first group are PIPS that 'act primarily by affecting the plant'. The group have in common some sort of defence mechanism such that a pest would be less able to attach to, penetrate, or successfully invade a plant.

Examples include:

- Some sort of structural barrier such as wax or plant hairs;
- Inactivation of or resistance to a pest or substance (*e.g.* toxin) produced by a pest;
- Creation of a deficiency in a nutrient or growth factor required by a pest;
- Hypersensitivity response that would contain or limit spread of the pest (*e.g.* necrosis of surrounding plant tissue, creating a functional barrier);
- Plant hormones that are naturally occurring but with change can affect a plant in a variety of ways with potentially dramatic results.

The other classes of PIPS proposed for exemption include viral proteins and their encoding genes. The expression of genes for viral proteins within a host plant confers resistance to infection by that virus as well as certain related viruses. These PIPS are plant virus specific, and as such, would not be expected to pose any risk to the public.

Herbicide-tolerant Crops

In the case of herbicide-tolerant crops, EPA determines whether the property of increased resistance may encourage higher rates or more extensive application of an agricultural chemical to food crops. The higher exposures to herbicides would be subject to evaluation through risk assessment, requiring the submission and review of detailed data. If the risk is determined to be acceptable, a herbicide product would require label extensions for the addition of new tolerant crop varieties.

Experimental Use Permit (EUP)

Small-scale research involving PIPS is covered under the permit and notification process that exists for new plant varieties administered by APHIS. The US EPA requires that an Experimental Use Permit (EUP) be obtained for all research and field testing involving greater than 10 acres of a plant modified to express a pesticidal substance. The EUP process allows a developer to gather the product use, performance, residue and other types of data, required in support of the registration of PIPS. All US EPA decisions concerning applications for EUPs are published within the Federal Register for comment.

Registration RMolecularMakers

For the registration of a plant-incorporated protectant, the EPA evaluates the potential impact, on human health and the environment, of the pesticide substance and of the genetic material necessary for its expression within the modified plant. With this in mind, the EPA developed their guidance for evaluating the safety of PIPS based on data requirements that were established for microbial pesticides. A number of modifications to the reporting requirements were required to focus the assessment on non-target toxicity rather than pathogenicity, as well as to evaluate the expression of the pesticide substance within the plant itself, rather than by spraying, etc. Although specific reporting requirements for the registration of PIPS are established on a case-by-case basis by the EPA, the general categories of information and data required in support of the registration of a plant-pesticide include:

- Product characterization
- Toxicology
- Environmental fate and exposure
- Non-target organism effects
- If necessary, plans for pest resistance management.

The EPA continues to hold meetings of their SAP to re-evaluate the reporting requirements that have been established for PIPS and to ensure that all potential human health and environmental concerns are being addressed as new scientific information becomes available. In addition, the developers of PIPS are encouraged to consult with US EPA during all phases of the research and development process, particularly when it concerns an environmental release.

As a part of their review, scientific evaluators within the EPA weigh the potential hazards inherent to the generic consideration of PIPS and the potential for exposure to the protectant substance. The characterization of possible hazards, in the context of exposure, allows for estimates of potential environmental risk to be established.

The commercial approval of a PIP may take between 15 and 18 months following receipt of a full data package by the US EPA, and it involves a public notification and consultation process. However, this time frame would be considered an ideal, since in reality, deciding what constitutes a 'full' data package may take several rounds of discussions with the US EPA.

6

Crop Ammoniation

COMPONENT OF STRAW AMMONIATION

The main component of straw is fibre, including cellulose and hemicellulose that can be digested by ruminants. Some cellulose and hemicellulose are bound to lignin and resistant to microbial attack. The role of ammoniation is to destroy this link, so these fractions are available to the animal.

Ammoniation usually increases digestibility by 20 per cent and CP content up to 1-2 times. It can also improve palatability and consumption rate. The total nutritional value can be doubled, reaching 0.4-0.5 feed units for each kilogram of ammoniated straw. In addition, ammoniation reduces mould development, destroys weed seeds (*e.g.,* wild oat, false sorghum, etc.), parasite eggs and bacteria.

AMMONIA SOURCES FOR STRAW AMMONIATION

The sources of ammonia to treat straw include anhydrous ammonia, urea, ammonium bicarbonate and aqueous ammonia.

Means of Anhydrous Ammonia

Anhydrous ammonia means "ammonia without water." Its formula is NH_3, and its N content is 28.3 per cent. The normal dosage is 3 per cent by weight of the straw DM. It is the most economical source of ammonia.The boiling point of anhydrous ammonia is -33.3°C, its vapour density is 0.59 (that of air is 1) and its liquid density 0.62 (that of water is 1).

Gas pressure is 1.1 kg/cm^2 at -17.8°C and 13.9 kg/cm^2 38°C. At normal temperature and pressure, anhydrous ammonia is a gas. Expensive pressure containers are required not only to keep it as a liquid, but also to transport and store it. Anhydrous ammonia is a potentially dangerous and toxic material, and stringent safety precautions need to be observed when using it. Its natural ignition temperature is 651°C. If the ammonia content in the air reaches 20 per cent, an explosion from self-ignition could occur. Attention should be paid to possible ammonia explosions, even though it seldom happens.

Urea

The N content of urea is 46.7 per cent. Its formula is $CO(NH_2)_2$. It is decomposed into ammonia and CO_2 by ureases at ambient temperature. The chemical reaction is:

$$CO(NH_2)_2 \cdot H_2O \xrightarrow{\text{Urease enzymes. Ambient Temperature}} 2NH_3 \cdot CO_2$$

Urea dosage needed to treat straw may vary a lot. The recommended dosage is 4-5 per cent urea on DM basis, taking into consideration the effect of ammoniation and costs. Urea can be transported conveniently at normal temperature and pressure. It is harmless to humans. Treating straw with urea does not need complex equipment and the sealing conditions are not as strict as with anhydrous ammonia. It is known that farmers in Bangladesh ammoniate straw in bamboo baskets lined and covered with leaves of a kind of banana.

From the *Yellow Cattle Magazine*, it is known that, in Anhui province, technicians use urea as a source of ammonia to treat straw without cover and get good results, which is beneficial for extension of straw ammoniation to rural areas. At present, urea is a widely used source of ammonia in China. Urea is not as effective as anhydrous ammonia for straw treatment, but it is better than ammonium bicarbonate.

Content of Ammonium Bicarbonate

The nitrogen content of ammonium bicarbonate is 15-17 per cent; its formula is NH_4HCO_3. It can be decomposed into NH_3, CO_2 and H_2O at a suitable temperature (above 60°C). The chemical reaction is:

$$NH_4HCO_3 \xrightarrow{\text{Heating}} NH_3 \cdot CO_2 \cdot H_2O$$

The dosage of ammonium bicarbonate, estimated by its N content, is 14-19 per cent of straw DM. But, according to the experiments carried out in Zhejiang and Shanxi Agricultural Universities, using 8-12 per cent of ammonium bicarbonate can give the same result as with 14-19 per cent.

Ammonium bicarbonate is a major product of the fertilizer industry and it is readily available at low price. It has a retail price of ¥ 300/ton (compared to more than ¥ 1000 for urea) and it is easy to use. Since ammonium bicarbonate is an intermediate product of urea breakdown, theoretically, in the right concentration, its effect should be similar to urea. Reports from Zhejiang Agricultural University indicate that in the humid south, straw ammoniated with urea showed more mould spots than with ammonium bicarbonate. It does not decompose completely at low temperature, thus in cold climates the effectiveness of treatment with ammonium bicarbonate is not good. When treating with ammonium bicarbonate in an oven, one day is enough, since the temperature reaches 90°C and it decomposes completely.

Aqueous Ammonia in Water

Aqueous ammonia is a solution of ammonia in water. The concentration is quite variable, but the usual value is 20 per cent. At this concentration, the normal dosage is 12 per cent by weight of straw DM. It is only adapted to areas near to fertilizer factories because its low N content makes transport expensive.

Other Sources

Besides the above sources of ammonia, human and animal urine also can be used to treat straw. However, collection difficulties limit practical applications.

Silo or Bunker Method

Ammoniating straw with urea in silos or bunkers has been widely used in China. This method has many advantages: silos or bunkers may be used for either ammoniating or ensiling year round; bunkers are easy to manage and avoid rodent damage to plastic films. Silos or bunkers constructed with cement are the best, since they save on plastic (only one sheet is needed, to cover) and minimize repairs. Once a silo or bunker is constructed, it can be used for several years. In addition, the bunker facilitates the estimation of straw weight.

Animal type and quantity determine bunker size. It should be known how much straw (air dried) can be placed per m^3 of bunker; how much ammoniated straw an animal requires per year; whether the bunker may also be used for ensiling; and how many treatments are required per year.

Average weight of air-dried and chopped crop residue (wheat, rice and maize), measured in different regions, is about 150 kg per m^3. Straw intake varies with animal type, size and concentrate level. Generally speaking, daily straw intake is 2 to 3 per cent of liveweight.

For example, daily net straw intake is 4-6 kg for a 200 kg steer, equivalent to 1.5-2 tonne per year. If the silo is used for straw, 1 m^3 will hold about 650 kg (25 per cent DM). With these data, the size of a silo or bunker can be designed according to practical needs (number of ammoniations per year, numbers of animals).

There are many types of bunkers. They can be built on the surface, underground, or half-and-half. It is recommended to build rectangular bunkers that, by adding internal walls, can be divided into double or twin bunkers for sequential treatment. If a double bunker is used for silage, a second fermentation can be started in the second compartment while straw is being used from the other half. Double bunkers are very common in Henan Zhoukou region.

For instance, a bunker with a volume of 2 m^3 requires 500 bricks, a bag of cement and a wagon of sand. Its total cost is ¥ 100. The cost of a double-bunker holding 4 m^3 is ¥ 200, affordable to farmers. One bunker of 2 m^3 can hold 300 kg of wheat straw. The two bunkers can be used in turns. One bunker of ammoniated straw is enough for two cattle for a month. If a bunker is used for silage, one bunker can hold more than 1 000 kg. The silo or bunker method has been well received by farmers.

Operation

First, straw is chopped to about 2 cm long. The general principle is that thick and hard residues, such as maize stover, should be cut shorter, while soft materials may be a little bit longer. Then urea (or ammonium bicarbonate) is added to water and stirred to completely dissolve it. Normally, 100 kg of dried straw needs 5 kg urea and 40-60 kg water.

Next, the urea solution is sprayed repeatedly over the straw. Before loading the silo, straw can be spread in an open area to facilitate uniformity in spray application. While straw is added to the silo, each layer should be compacted till the bunker is full, and then it is covered with plastic film, held firmly in place by a layer of fine soil. Treatment time with urea is a little longer than with anhydrous ammonia.

When treating straw with urea, the speed of urea decomposition into ammonia should be taken into consideration. It depends on ambient temperature and amount of ureases present in the straw. Decomposition rate increases with temperature, so this method is well adapted to warm regions (or warm seasons). In general, it is suggested to add some substance rich in ureases, such as soybean cake powder, to accelerate urea breakdown.

Maize stover contains much more ureases than other cereal straws, so ureases are not required. In China, maize stover is one of the three main crop residues, along with rice and wheat straw. The nutritive value of maize stover is higher than other straws, either before or after ammoniation. However, it was seldom used as feed because it is thick and hard, and not easy to be store and transport. Nowadays it should be preferentially used as feed because processing and treatment methods (chopping, kneading, heat-extrusion, baling and ammonia treatment) are available to overcome these problems.

Oven Method

Straw can be quickly treated quickly with ammonia in an oven. The Non-conventional Feed Institute of China Agricultural University has developed a metal, self-assembly oven for ammonia treatment. The oven is composed of a chamber, a heating system, an air circulating system and a straw trailer. The chamber must be insulated, sealed and resistant to corrosion by acid or alkali. The heating device may be an electric heater or a coal-fired steam heater, depending on local conditions. Shanxi Wanrong County used steam to heat the oven, with good results. The straw trailer should be convenient for loading, unloading, transport and heating. A metal mesh trailer with steel wheels is the preferred option.

Operation

Ammonium bicarbonate (8-12 per cent of straw DM) is first dissolved in water. Straw (baled or loose; chopped or whole) is placed on the trailer. The solution is uniformly sprayed over the straw, adjusting moisture content to about 45 per cent. Once the trailer is full, it is moved into the chamber, the door is closed and heating started. If heating is by electricity, the heating tube needs to be set to control the chamber's temperature at about 95°C. After heating for 14-15 hours, it is turned off. The chamber should remain sealed for 5-6 hours more, and then the straw trailer is moved outside for ventilation. Once ammonia residues are gone, the straw can be fed.

The oven can shorten treatment time considerably since it only takes 24 hours to treat straw. In addition, oven treatment is not weather dependent, so ammoniation can be done all year round. However, the cost is relatively high, limiting its application. In recent years in Jilin Province, straw has been treated in a coal-heated chamber usually used for flue-curing tobacco. The investment for the oven was avoided and fuel expenses reduced, so treatment cost was greatly decreased.

Other Options

Besides the above three widely used methods, there are still some other methods for straw treatment. For instance, ammoniation within plastic bags was readily accepted by farmers due to low initial cost. However, repeated purchases of plastic bags increased costs and restricted wider application. In some places, straw is treated in locally available containers (such as vats). This latter method may be worth advocating.

AMMONIATING OF LOW QUALITY FORAGES

Ammoniating low quality forages is a management technique that can be used to treat forages like wheat and oat straw and baled cornstalks. It is not recommended to be used on grass hay, summer annuals, and alfalfa.

Toxicity can occur when medium quality forages are ammoniated and fed to cows and the calves who's dam consume ammonia treated forages. The toxic compound is transferred through the milk to the calf. Affected calves walk in circles and this is commonly referred to as circling disease. If toxicity occurs in the calves or cows, avoid working or moving the cattle and remove the forage. To my knowledge, circling disease has not occurred in cattle as a result of consuming ammoniated straws or crop residues.

Treating wheat straw with anhydrous ammonia can make straw almost as digestible as average quality prairie hay. Ammoniating will increase digestibility of low quality forages and, therefore, intake will increase. Cattle don't quit eating straw because they don't like it; but, because of its low digestibility and slow rate of passage, they can't stuff anymore into their rumen. The ammoniation process is temperature dependent and occurs faster at higher environmental temperatures.If the temperature is 86EF, the ammoniated residue needs to be kept sealed for one week. If the temperature is between 59EF and 86EF it needs to remain sealed for 2 to 4 weeks. If the temperature is below 59EF, the ammoniated forage needs to be sealed for 4 to 8 weeks. It is important to keep the package sealed and not let the ammonia escape. It can be difficult to keep the package sealed for a long period of time because of wind and curious pets and wild animals.

For residues like wheat straw, bale the straw soon after grain harvest, preferably with some moisture on it or bale early in the morning when there is some dew present. To prepare an area to ammoniate, scrap an area that will accommodate three big round bales wide with 5 to 6 feet on either side and in the front and back and 10 to 12 bales long. Push the dirt to the sides of the area where the bales will be stacked. The dirt will be used to seal the plastic around the edges.

Gather bales into rows that are stacked in a pyramid (three bales on the base, two bales on the second level, one bale on the top or three bales on the base and two bales on the second level) leaving a couple of inches between pyramids for the ammonia to filter around the bales. Cover the entire stack with one sheet of 6 to 8 millimeter black plastic. If the plastic is 40 by 100 feet, you will be able to cover 10 to 12 pyramids in a row. Make sure the edges of plastic on the ground are sealed with loose soil to prevent leaking of ammonia. Any holes in the plastic can be patched using duct-tape.

Place a pipe (6 to 8 feet long) on the ground and insert it into the center of your stack. Next, attach the pipe to the anhydrous tank and slowly leak anhydrous into the bales sealed with plastic. Ammonia can be dangerous so be careful when working with this product. Don't inject ammonia too fast or the plastic can rupture. Continue to add anhydrous slowly until you have added 60 pounds per ton of residue on a dry matter. This process will take about 10 minutes for each ton of forage ammoniated. When completed, turn off the tank, remove the pipe, and seal its opening with dirt. Keep the stack sealed to allow for the reaction to be completed. About a week before feeding, open one end of the stack to allow excess ammonia gas to escape. If you sample treated straw and conduct a nutrient analysis, the crude protein content of the treated straw will be greater than the non-treated straw. To measure protein content of a feed/forage, the analysis measures the NH3 groups present in the sample and because NH3 groups were added through the ammoniating process, it stands to reason that crude protein content of the treated straw be higher. Some of the NH3 in the treated straw will be used by bacteria in the rumen to make bacterial protein.

More importantly than the increase in crude protein content is the increase in digestibility by about 10% of the treated straw compared to the untreated straw. The increase in digestibility of the treated residue allows cattle to consume more of the treated residue compared to non-treated straw by about 20%. This allows the animal to meet a greater portion of their nutrient needs by consuming the treated forage.

QUALITY OF SILAGE

The nutritive value and the quality of silage should be accurately evaluated. Working on behalf of the Bureau of Animal Production and Health (BAPH), MOA, researchers at Zhejiang University drafted methods of evaluating nutritive value and quality of silage, which have been tested in China since 1996. This handbook includes subjective methods (on-farm) and chemical methods (for use in laboratory).

SUBJECTIVE METHODS OF EVALUATION

The pH value and certain simple subjective criteria such as colour, smell and texture are used to evaluate the quality of silage on-farm.

Colour

Good silage usually preserves well the original colour of the standing plant. When green raw material produces silage with green or yellow colour, it can be considered of good quality. Temperature is one of the important factors affecting silage colour. The lower the temperature during ensilage, the less colour change. Above 30°C, grass silage becomes dark yellow. Above 45 to 60°C, the colour becomes closer to brown. Beyond 60°C, the colour darkens towards black due to caramelization of sugars in the forage.

However, silage quality can be misjudged by on a colour basis. For example, silage from red clover or Chinese milk vetch is often dark brown instead of light brown. Despite its excellent quality, it may be considered a failure due to colour. A more useful indicator is colour of the silage juices. It can generally be said that the lighter the colour of the juice, the greater the success.

pH Value

The pH is the simplest and quickest way of evaluating silage quality, and may be determined on-farm using wide-range pH test papers such as bromophenol blue (range 2.8-4.4), bromocresol green (range 4.2-5.6) and methyl red (range 5.4-7.0). The classification of silage based on pH value is:

pH below 4.0	Excellent
pH between 4.1 and 4.3	Good
pH between 4.4 and 5.0	Average
pH above 5.0	Bad

Smell

Good silage usually has a mild, slightly acidic and fruity smell, resembling that of cut bread and of tobacco (due to the lactic acid). A rancid and nauseous smell denotes the presence of butyric acid and signifies a failed silage. A musty smell is a sign of deficient compaction and presence of oxygen. A distinctive unpleasant smell, of sow's urine and faecal matter, is always indicative of marked protein degradation during ensilage.

Texture

Plant structures (stems and leaves) should be completely recognizable in the silage. A destroyed structure is a sign of severe putrefaction. A viscous, slimy appearance reveals the activity of pectolytic (sporulating) micro-organisms.

Taste

This test is more suitable for specialists. It is of little value to the farmer, whose basis is the palatability to farm animals.

EVALUATION OF SILAGE IN THE LABORATORY

Evaluation of silage in the laboratory is mainly based on chemical analysis. It includes pH determination and assay of organic acids (acetic, propionic, butyric and lactic), which are the main fermentation metabolites in silage. Fermentation characteristics may be estimated based on total acids or on proportions of individual acids. Free ammonia assay, and better still the ratio of ammonia nitrogen to total nitrogen, is the most valid criterion for protein degradation.

Before discussing the value of such analyses, it is necessary to emphasize the necessity for the sample to be truly representative. Samples must be taken with a probe from several places at different layers in the silos. The sampling points should be more than 30 cm from the edge of the silo to prevent misleading results. Samples should preferably be sent to the laboratory in sealed plastic bags or glass bottles to avoid aerobic deterioration (secondary fermentation). Ammonia nitrogen and pH should be determined as soon as possible.

Silage Evaluation by pH

The pH of silage should be measured in the laboratory using a precise pH meter. For fresh grass silage (moisture above 75 per cent) and maize silage at all DM contents, the pH is both the simplest and quickest method of evaluation. Research has shown that there is a very close relationship between pH value, fermentation quality and DM during ensiling. The criteria for evaluation of silage from pH values.

Silage Evaluation by the Ratio of Ammonia-N to Total N

The ratio of ammonia-N to total N (NH_3-N/TN) in silage provides information on the stage of protein degradation and it undeniably constitutes a test of the state of conservation of the ensiled proteins.

The higher the ratio, the more protein has been degraded, and the worse the quality. In the system proposed, a maximum of 50 points is given for a ratio lower than 5 per cent, and points are deducted for a ratio higher than 35 per cent. Table presents the point scale for the evaluation of silage from ammonia N.

Table. Scale for Silage Evaluation from the Ratio ofAmmonia-N to total N

Silage Quality	% NH_3-N/TN	Points	Silage Quality	% NH_3-N/TN	Points
Very good	<5	50	Average	15.1-16.0	22
Good	5.1-6.0	48		16.1-17.0	19
	6.1-7.0	46		17.1-18.0	16
	7.1-8.0	44		18.1-19.0	13
	8.1-9.0	42		19.1-20.0	10
	9.1-10.0	40	Bad	20.1-22.0	8
Satisfactory	10.1-11.0	37		22.1-26.0	5
	11.1-12.0	34		26.1-30.0	2
	12.1-13.0	31	Very bad	30.1-35.0	0
	13.1-14.0	28		35.1-40.0	-5
	14.1-15.0	25		>40.1	-10

Silage Evaluation using Various Organic Acids

The presence of lactic acid and various volatile fatty acids, especially acetic, propionic and butyric acids, is a reflection of the fermentation that has occurred. It is appropriate therefore to take into account, when judging silage success, both the type of acids and the amount present. The higher the proportion of lactic acid, the better the quality.

The evaluation system is basically that of Flieg from 1938 with a maximum of 100 points given.

Full points are given for lactic acid above 68 per cent, for acetic acid below 20 per cent and for butyric acid below 0.1 per cent. Individual acids are scored independently, and the total is the sum of the three acids. Table presents the key for silage evaluation according to Flieg (0 - 20 = failure; 21 - 40 = poor; 41 - 60 = satisfactory; 61 - 80 = good; 81 - 100 = very good).

Integrative Silage Evaluation from Ammonia-N and Organic Acids

In order to integrate the information from both protein and carbohydrate in the silage, an evaluation system based on the points from ammonia-N and organic acids is proposed. There is a maximum of 100 points in the scale, 50 points for proteins and 50 points for carbohydrates. Points for carbohydrates are obtained by dividing the data in Table. The following table is used for evaluation of silage quality from protein and carbohydrate:

Total points	0-20	20-40	40-60	60-80	80-100
Overall quality	Very bad	Bad	Average	Good	Very good

FEEDING SILAGE

Silage is suitable for feeding some 6-7 weeks after ensiling. The basic principle for silage use is to minimize the area exposed to air, with as little churning as possible. Silage should be taken from one side of the trench, and the exposed surface should be immediately covered to minimize re-entry of air and avoid secondary fermentations. Material taken out from the silo should be fed as soon as possible, since it rapidly deteriorates upon contact with air. Silage left in the trough must be cleaned away in time to prevent putrefaction.Some animals may not like silage when offered for the first time. In these cases, some adaptive measures should be taken. For example, silage can be placed in the bottom of the trough, and covered with concentrates. When animals are adapted to silage, the offer can be increased.

Silage can be fed to all kinds of animals: replacement cattle, fattening cattle, dairy cows, sheep, goats and even pigs. The amount of silage offered depends on the animal and its age, as well as on the type and quality of silage. Taking cattle as an example, the recommended amount to be fed daily is:

Grass Silage	4 kg per 100 kg Liveweight,
Legume Silage	3 kg per 100 kg Liveweight,
Maize Stover Silage	4 kg per 100 kg Liveweight,
Starch-rich Forage (Whole-maize Silage)	5 kg per 100 kg Liveweight,

Therefore a cow with liveweight of 500 kg may be daily fed 20 kg maize stover silage.Sometimes the milk from cows fed on silage has a taint. Because this smell is only transmitted through the air, the following points should be observed to prevent it:

- *Handling:* Silage should never remain in the cowshed, and it should be offered in amounts exactly as required.
- *Feeding method:* Silage should not be given before milking. The effect on milk taste is more marked when the silage is fed 2 hours before milking, and least when given 6 hours after.
- *Hygiene and cleanliness:* Both floor and operators should be clean and the shed well ventilated.
- *Milk and milking:* Equipment should be kept clean, and milk cooled as soon as possible.

It is sometimes believed that ensiled forage has an adverse effect on the general health of the animal, especially on the skeleton of young animals. This criticism has no scientific evidence. It is certain that musty silage adversely affects animals, but such material should never be fed in the first place. It must be pointed out that silage is a good feed, but it is certainly not a "complete feed." For example, maize silage contains insufficient Ca and P, and should be supplemented. P content in ensiled alfalfa is sufficient for heifer growth and Ca more than enough. Animals fed on silage should be properly supplemented with the nutrients that are insufficient for the animals' requirements.

It must also be emphasized that the nutritional value of silage depends on the quality of the original forage, harvest time, conservation method, etc. Nutritional values of different silages therefore vary substantially. When it is impossible to analyse silage at the laboratory, one should refer to feed manuals and roughly assess the nutritional value of the silage to be fed, in order to formulate a ration that matches the animal's requirement.

TYPES OF SILOS

STACK SILO

This type of silo implies a pile of material on the ground surface. On flat and dry ground, plastic sheet is placed underneath and the material is laid in a stack. The top is covered with plastic and sealed all round with soil. Sandbags or old tyres, or any other suitable objects, are placed on top to prevent the top cover from being blown away by the wind. The advantages of the stack silo are low cost and flexibility of placement.

PLASTIC SILO

Animal scientists from Neijiang, Sichuan Province, successfully ensiled sweet potato vines in plastic bags in 1978.

The plastic silo is similar to the stack silo but it is covered with plastic sheets of polyvinyl chloride (PVC) or polyethylene. Alternately, the silo can be made in bags with sealed tops. The stack silo is also inexpensive and can be placed anywhere. However, labour requirements are high due to manual filling and handling. A specific machine has recently been developed by the Grassland Institute, Chinese Academy of Agricultural Science, to fill plastic bags with forage.

TOWER SILO

In China, tower silos are constructed from brick, and are several metres in diameter and 10-20 m in height. The advantages of this type of silo include: long life, small space required, low storage losses, and possibility for mechanization. Both the filling operation and daily extraction can be mechanized. However, tower silos are expensive, and therefore not widely used in China, with the exception of some state-owned farms.

CELLAR SILO

The cellar type is the most common silo on individual farms. Round or square concrete silos are usually built inside houses for protection from the weather. Advantages are lower cost and easy management. Size can be adjusted according to scale of production. Cellar silos are suitable for rural conditions in China. A disadvantage is high effluent loss, especially with clay walls.

TRENCH SILO

This type is generally built underground or semi-underground, with two solid walls of 1.5-2 m in height. Advantages are similar to the cellar silo, but the trench silo is more suitable for mechanization. The tractor can be driven on top from one side to the other for compaction purposes. After compaction, it is covered with a plastic sheet pressed down with soil, sandbags or straw bales to maintain anaerobic conditions.

On many dairy farms, trench silos are built on the surface of ground. This type of trench silo resembles a bunker silo, but has vertical walls of 0.4-0.5 m in thickness and 3-4 m in height. This design makes mechanization more convenient, and may also prevent bottom leakage.

CONTROL OF MOISTURE CONTENT IN RAW MATERIALS

Ensilage can only be successful, with minimum DM and nutrient losses, when the moisture content of the raw material is kept to a suitable level. Although silage may be made within a large range of moisture contents, DM should be over 20 per cent to assure silage quality.

There are many disadvantages to ensiling crops with high moisture content. First, ensiling of wet materials results in the generation of a large volume of effluent, which not only poses disposal problems, but also carries off valuable, highly digestible nutrients in solution. The amount of effluent increases with silo height, due to pressure. Effluent is produced when moisture is above 75 per cent. Secondly, the critical pH value for clostridial growth varies directly with the moisture content of the plant material, and unless soluble carbohydrate levels are exceptionally high, ensiling wet crops will encourage clostridial fermentation, resulting in high losses and reduced nutritive value. Thirdly, even if the water-soluble carbohydrate (WSC) levels are adequate to ensure lactic fermentation, very wet silages may still be nutritionally undesirable because voluntary DM intake of these is frequently low. Finally, drier plant materials are preferred because they are easier to handle and a higher quantity of DM can be carried per trailer load.

Moisture content of forage and grasses is above 80 per cent when harvested at a suitable stage. Therefore the moisture content should be reduced by field wilting. It takes 4 to 6 hours to wilt in dry regions such as the north western provinces and Inner Mongolia, and 6 to 10 hours in north eastern and northern areas. Longer periods may be needed in southern provinces, depending on climate and weather, but it is not desirable to exceed 24 hours. When weather conditions are unfavourable, field wilting should be avoided to prevent nutrient loss due to rain leaching. In these cases, other methods should be considered.

In contrast, the moisture content of cereal straw is generally too low to allow tight packing, so cereal straw and stover should be finely chopped. Sometimes water should be added to bring moisture content to a suitable level.Sweet potato vines are high in moisture content and wilting is necessary. Chopped vines are usually mixed with finely chopped straw or bran prior to ensiling to increase overall DM content. The moisture content of plant materials may be measured with instruments, but it is usually estimated manually on farm.

Samples of chopped and minced grass or leguminous forage are grasped tightly by hand for one minute or so to estimate moisture content. If juice can be extracted, moisture content is above 75 per cent.

If the material remains together but without juice, moisture content is between 70-75 per cent. If the material has elasticity and spreads out slowly, moisture content is 55-65 per cent. If the material spreads out quickly, moisture content is about 55 per cent. If the material breaks, moisture content may be below 55 per cent.

CHOPPING, COMPACTION AND SEALING

Prior to ensiling, plant materials should be chopped. The fineness of chopping varies with moisture content and nature of the material. The following guidelines can be used, but, in principle, rough and hard materials should be finely chopped, while delicate and soft materials can be roughly chopped.

High moisture forage (Moisture >75%)	chop to 6.5-25 mm
Wilted materials (Moisture 60-70%)	chop to 6.5 mm
Whole maize plant	chop to 6.5-13 mm

What are the advantages of chopping? Firstly, chopping facilitates compaction and thus reaching the anaerobic stage. When most oxygen is removed, clostridial growth is discouraged and lactic acid fermentation encouraged. Secondly, chopping releases plant juices, stimulating the growth of lactic acid bacteria. Thirdly, chopping may increase silage intake by improving quality of fermentation and by accelerating rate of passage of feed particles through the rumen. However, very finely chopped silage reduces the rumination and may decrease milk fat content. Thus, 10-15 per cent of the silage material should be above 25 mm in length in order to maintain an effective fibre function.

FACTORS INFLUENCING SILAGE QUALITY

Quality of silage fermentation is influenced by several factors, including moisture, WSC content of raw materials, degree of compaction and effectiveness of final sealing. Stage of maturity influences a forage's nutritive value and thus quality of silage. Leguminous forages at budding stage have optimal energy, protein and carotene contents, but DM yield and nutritive value decrease with maturity. Grass at heading stage is highest in caloric value and protein content. When leguminous forages reach late flowering stage or grasses reach seed stage, the energy value has decreased by 70-75 per cent, digestible CP by 67-83 per cent, and carotene content by 75-84 per cent.

Therefore, it is important to harvest forage crops at their optimal stages. Legumes should be harvested at budding stage, grasses at heading stage, and whole maize plant at late milky to early dough seed stage.

Grass is highest in protein at pre-heading, and highest energetically at dough seed stage. Many grasses can not head again after the first cutting, and the re-growth may be collected 4 to 5 weeks after the first harvest. The appropriate stages at which plants may be harvested are indicated in Table.

Because straw and stover are harvested after grain collection, their nutritional value is generally low. It has been recently shown that yield and grain quality do not change when harvest is shifted to an earlier stage (by 7-10 days). However, this shift may be beneficial in terms of improved nutritional value of straw and stover.

SILAGE ADDITIVES

The purpose of using additives is to ensure silage quality by encouraging lactic acid fermentation, by inhibiting undesirable microbes or by improving its nutritional value. Common silage additives include bacterial cultures, acids, inhibitors of aerobic damage, and nutrients.

BACTERIAL CULTURES

A dominant lactic acid fermentation is the key to making good silage. Lactic acid bacteria are normally present on harvested crops together with clostridial bacteria in a ratio of about 10 to 1. Considerable nutrient loss usually occurs at initial stages of ensiling when oxygen is still present. Addition of lactic acid bacterial cultures during filling-up increases their population rapidly, encouraging lactic acid fermentation and pH reduction to a level that inhibits clostridial development. Different strains of lactic acid bacteria look similar under the microscope, but their biological activities are very different.

Only those acid-tolerant strains that possess a homo-fermentative pathway, producing the maximum amount of lactic acid from hexose sugars readily available, and a growth temperature range extending to 50°C, should be used as a silage additive. The environment under which lactic acid bacteria multiply favourably is also important. Lactic acid bacteria are anaerobic, and hence air should be removed and the silo should be kept airtight. These micro-organisms ferment soluble sugars to a mixture of acids, but predominantly lactic acid.

Plant materials should contain at least 2 per cent WSC, otherwise soluble sugars (*e.g.*, molasses) should be added. Starch may also be added along with amylase in order to provide lactic acid bacteria with soluble sugars.Cereal straws and stovers are high in lignocellulose. A mixture of inoculum or enzymes, or both, containing cellulase and xylanase is often used. Enzymes degrade lignocellulose, liberating soluble sugars for bacterial use. A number of commercial preparations are available from foreign companies, and some have been registered by government authorities and can be sold in China. During the Ninth Five-Year Plan, Chinese scientists successful produced a special additive for ensiling fresh cereal straws and stovers.

MINERAL OR ORGANIC ACIDS

The original proposal to use acids as silage additives dates back to 1885. In the late 1920s, Virtanen from Finland, adopted this approach and recommended the rapid acidification of the crop with mineral acids (AIV process) to a pH of about 3.5, which was originally thought to inhibit microbial and plant enzyme activity. This AIV process was widely used in Scandinavia until quite recently. Due to the difficulties in handling corrosive acids, organic acids were later used as silage additives. When acids are added, plant materials sink quickly and are easy to consolidate. Acidity may arrest plant respiration and reduce heat production and nutrient loss. Rapid acidification may also inhibit clostridia. However, addition of acids increases effluent and can be potentially toxic to animals. Furthermore, acids are corrosive to people, animals and machinery. Reduction of moisture content may minimize effluent, and addition of calcium carbonate can be used to adjust silage acidity.

Appropriate concentrations of different acids as silage additives are recommended as follows:

- Sulphuric and hydrochloric acids: 50-80 litre of dilute acid (concentrated acid diluted with water at 1:5 v/v) per tonne
- Formic acid: about 3 kg/ton
- Acetic acid: 5-20 kg/ton
- Propionic acid: 1 litre/m^2 of surface to prevent mould development.

INHIBITORS OF AEROBIC DETERIORATION

The most common inhibitors of aerobic deterioration are sodium nitrate, sodium nitrite, sodium formate and formaldehyde. These chemicals do not contribute to the improvement of fermentation, but are effective in preventing silage deterioration. Some plant parts, such as larch leaves, contain natural bactericides that may safely function as antiseptics. Formaldehyde is a well-known sterilizing agent and is commercially available as formalin, which contains 40 per cent of the gas in aqueous solution. Scientists are interested in the additives because of their bacteriostatic properties, and because of their known properties of protecting plant proteins from rumen microbial degradation. Adding formaldehyde at 3-15 kg/ton may result in a well-preserved silage.

NUTRIENTS

Nutrient additives are defined as substances which, when added to ensilage materials, contribute significantly to the nutritional value of the silage. Most of nutrient additives can also favour lactic acid fermentation. A number of materials are considered to be in this category.

Nitrogenous Compounds

Certain crops, such as maize and most cereal straw, are nutritionally deficient in nitrogen, and when fed to ruminants as silage require supplementing with a protein supplement. An alternative approach is to improve the CP content of the silage by adding urea during ensiling. When applied to maize, urea produces silages with higher pH values and fermentation acid contents than in untreated silages.

Urea addition has also a marked effect on nitrogenous components of the silages, resulting in higher CP, true protein, free amino acids and ammonia. Addition of urea to cereal straw and stover may also have an ammoniation effect, which is associated with higher CP and lower fibre content. However, attention should be paid to the rapid release of ammonia from urea in the rumen. High concentrations of ammonia in the rumen may cause ammonia poisoning. One solution is to add the urea or other nitrogenous compounds together with soluble sugar sources, such as molasses and starchy grains.

Urea may be added at 0.5 per cent of fresh materials. When whole maize plant was added with urea at 0.5 per cent, the CP of the resulting silage increased to 12.9 from 8.7 per cent in the untreated silage. Urea may also be added prior to feeding the animals.

Carbohydrate-rich Materials

Carbohydrate-rich materials are added to silage crops in order to increase the supply of available energy for the growth of lactic acid bacteria, and are of particular importance in crops such as legumes, which are deficient in soluble carbohydrate content. Materials that have been used for this purpose include molasses and cereals. Molasses is a by-product of the sugar industry, and has a DM content of 70-75 per cent and a soluble carbohydrate content of about 65 per cent of DM.

In order to obtain maximum benefit, it should be used at 4 per cent (w/w) for grass silage and 6 per cent (w/w) for legume silage. Cereals have been used as additives in an attempt to improve both the fermentation quality and the nutritional value of silages. Cereals contain 50-55 per cent of starch that can be hydrolysed to soluble sugars and then utilized by lactic acid bacteria. If amylase or amylase-rich materials, such as malt, are added with the cereal to ensiled crops, they will be more effective as fermentable carbohydrate sources.

Minerals

In addition to being nutritionally deficient in N, most straw and stover is a poor source of Ca and many micro-elements. Limestone is sometimes added to silages as Ca supplement and to alleviate silage acidity. Calcium carbonate may be added at 0.45-0.50 per cent to obtain maximum benefit. Common salt has a high osmotic pressure to which clostridia are sensitive, but lactic acid bacteria are not. Addition of salt may increase lactate content, decrease acetate and butyrate, resulting in silage with good quality and palatability.

Other minerals may be used in ensiled materials. Examples of mineral sources include copper sulphate (2.5 ppm), manganese sulphate (5 ppm), zinc sulphate (2 ppm), cobalt chloride (1 ppm) and potassium iodide (0.1 ppm).

Finally, it has to be pointed out that recommendations on the use of silage additives must be based not only on the results of scientific research, but also on sound economic return.

EFFECTS OF SUPPLEMENTATION WITH GREEN FORAGE

It is known that small quantities green forage can improve the usage of straw diets. Thus, introducing forage supplements may be an alternative strategy for increasing nutrient intake and improving ruminant performance. A wide range of forage supplements is available in China, depending on location. These include green forage, crop by-products and aquatic weeds.

Chinese milk vetch (*Astragalus sinicus* L.) is cultivated in South China as green manure to improve soil fertility. Farmers traditionally offer surplus vetch to swine after ensiling. Liu *et al.* and Ye *et al.* evaluated the effect of supplementing with milk vetch silage on growth rate of heifers and rumen function in sheep given ammoniated rice straw diets. Intake of ammoniated straw by heifers was slightly decreased as the level of vetch silage went up. When the vetch silage represented 20 or 30 per cent of diet, growth rate was significantly higher than in the non-supplemented group ($P < 0.05$). The highest gain was obtained at the 20 per cent level, when concentrate consumption per kilogram weight gain was lowest (1.25 kg).

The average ammonia-N level in the rumen increased with the increasing level of vetch silage. However, there were small differences in pH value and volatile fatty acid profiles among all groups. Protozoa population tended to decrease more quickly with the supplement. The microbial protein concentrations in the rumen fluid was related to the levels of vetch silage and reached a peak at 20 per cent, possibly associated with saponins present in milk vetch.Liu *et al.* (2001) substituted RSM with mulberry (*Morus alba*) leaves as a supplement for Huzhou growing lambs fed ammoniated rice straw. Total substitution with mulberry leaves gave similar growth rates, but at lower cost than RSM. The authors concluded that mulberry leaves could be used to supplement ammoniated straw diets in place of RSM.

Forage supplement level would depend on the type used, availability and the cost relative to straw. A maximum of 20-25 per cent of a diet seems suitable. The substitution rate would increase with the level of supplement. A limitation in China is that supplementary forages are usually in short supply in most areas.

SUPPLEMENTATION WITH READILY DIGESTIBLE FIBRE

Prerequisites for all ruminal digestive processes are development of digestive consortia and their adhesion to ingested feed particles. Any feeding strategies that enhance adhesion of rumen microbes to feed particles and improve fibrolytic activity may be beneficial to feed utilization.

In a recent trial, Shi *et al.* studied the effect of added ammoniated rice straw on the growth rate of Holstein heifers receiving untreated rice straw.

When half (w/w) of the untreated straw was replaced by ammoniated straw, heifers had significantly higher intake and gains, even slightly higher than those on ammoniated straw. A similar result was reported by Li *et al.* who compared the growth rate of cross-bred cattle offered ammoniated wheat straw or maize stover, either alone or in a 50:50 (w/w) combination.

Bamboo cultivation is very popular in south China. Bamboo shoot shells (BSS) are the residue from industrial-scale processing of bamboo shoots, and represent a disposal problem because they have no use and can pollute the environment. Occasionally it has been observed that fresh BSS were palatable to cattle, their CP was 10-13 per cent (on DM basis) and were easily degraded in the rumen, although neutral detergent fibre was high (65-70 per cent on DM basis).

Considering BSS as a source of readily digestible fibre, Liu *et al.* observed the response in growth rate to supplementation in heifers given ammoniated rice straw. Straw intake linearly decreased with the increasing level of BSS, but total dietary intake increased also (a substitution rate less than 1.0). Growth rate in heifers was improved significantly by the supplementation, and the optimum level was at 21 per cent of total dietary intake.It is inferred that supplementation with readily digestible fibre may improve utilization of basal diet and animal performance. However, further work is needed on this topic.

USE OF MULTINUTRIENT BLOCKS

Since the early 1980s, both production and utilization of multinutrient blocks (MNBs) as supplements for ruminant animals have increased considerably in developing countries. With the development of ruminant production, much progress has been made and new technologies have been developed in the manufacturing of MNBs in China since they were introduced in the late 1980s.

MNB Manufacture

Ingredients

The MNBs developed in China contain molasses, urea, minerals and proteins, with the aim of supplementing cereal straw with fermentable N, soluble carbohydrates, minerals and other nutrients. The main ingredients have been: ground maize; rice bran and wheat bran; rapeseed meal; solidifiers and binders (cement, clay, etc.); bone meal; and vitamin premix. Molasses, a source of easily fermentable carbohydrates and a binder, makes blocks highly palatable.

It has been demonstrated that mixing urea with molasses greatly decreases the release of ammonia-N in the rumen. Mineral premix usually contains Ca, P and Na as well as micro-elements such as Fe, Cu, Mn, Zn, I, Se and Co.In a series of demonstration trials in Gansu Province, where the basic diet was composed of wheat straw and other stubble, Chen *et al.* selected three formulas, for cows, heifers and calves. Many workers used molasses as an ingredient for blocks found a block formula without molasses, since molasses is expensive and in short supply in some regions.

Lime and cement have been commonly used as solidifiers and binders. Ordinary clay has been also proved to be efficient for making blocks. Farmers in some regions used loess as a binder.

Table. Formulas of Multinutrient Blocks for Dairy Cattle (on a Percentage by Weight Basis)

Ingredients	Cow	Heifer	Calf
Molasses	8	10	15
Urea	16	12	–
Salt	26	26	22.8
Ground maize	5	5	10
Lime	10	10	10
Clay	11.2	15	15
Bone meal	–	–	5
Mineral mixture	23.8	22	22.2
Total	100	100	100

Process for Block Manufacture and Specifications

Depending on the technical process, MNB preparation takes two forms: by pressure (hot process), or by moulding (cold process). The moulding process needs neither sophisticated equipment nor much energy.

Blocks produced by moulding had the following features:

- When water was poured onto the surface of the blocks, blocks kept their shape after sun-drying;
- Blocks maintained their shape intact when submerged in water for 1-2 hours, but completely dispersed after 4-5 hours;
- Hardness increased when formaldehyde was included;

- The shape of the blocks did not change under finger pressure.

Xia *et al*. developed a specialized machine to make blocks under pressure. This equipment saved space and labour, and the blocks could be easily produced. Drying was unnecessary since it used raw materials already dry. The blocks produced were compact, not deliquescent, and hard enough to control intake. They did not become mouldy nor did they lose shape when exposed to rain or sunshine.

The characteristics of two press machines for the formation of blocks, designed by Chen *et al*.. The blocks were made by mixing molasses and urea, and then heating. Salt was added, followed by the rest of the ingredients, having been previously mixed together. The complete mixture was then pressed and the resulting blocks were wrapped immediately.

Blocks made using both press machines were hard enough, with a breaking strength of 44 kg/cm^2. The blocks were oblate, 25.6 cm in diameter and 8 cm in thickness, with a weight of 7.5 kg each. The blocks produced by Yang *et al*. were squares or compressed cylindroids with rounded holes (ca 1.5 cm diameter) in the centre. Each block weighed 2.5 kg. The breaking strength was 56.9 kg/cm^2. They did not moisten before 24 hour under low temperature (> 0°C) and high humidity (> 80 per cent).The urea-mineral blocks designed by Liu *et al*. had a breaking strength of 40 kg/cm^2. They were easily transported and fed to the animals. Even in situations of high humidity, there were no losses from mould or from hydration when they were offered to the animals over a long period.

In 1999, MOA set up a "block expert group." Based on results of previous studies and six months of research, an industrial production technology system was proposed. The MOA appraised it, and confirmed it "in the national leading position." The blocks have sold very well.

GROWTH OF BLOCKS WITH ANIMALS

Beef Cattle

In a growth trial with heifers, those having access to MNBs had daily gains of 835 g/day, 112 g/day higher (P<0.05) than the control group. Animals supplemented with blocks reached 380 kg body weight (weight at first service) 65 days earlier. Other advantages observed during animal feeding trials on farms were better skin coat, better body condition, and lower death rate. The urea-MNBs without molasses were also palatable to both cattle and goats. Local Yellow cattle on grazings with access to blocks performed better than the control (370 *vs* 203 g/day). The animals with blocks had better body condition and looked healthier than the control group. An increased income of ¥ 0.57 could be obtained per beast per day.

Zhang *et al*. observed that daily gains were 15.6 per cent higher, and consumption of roughage and concentrate per kilogram of gain were 16.9 and

13.3 per cent lower, respectively, when beef cattle were supplemented with MNBs containing non-protein nitrogenous compounds (NPN). In another trial (Ma *et al.*, 1995), beef cattle with access to blocks containing NPN had daily gains 353 g higher than those with no blocks (1478 *vs* 1125 g/day).

Dairy Cows

Dairy cows supplemented with MNBs produced 1.06-1.47 kg (5.3-5.9 per cent) more milk than those without blocks. Less metabolic disorders occurred in the supplemented animals. Increased net income attributed to the blocks was about ¥ 1 per cow per day. Chen *et al.* found that cows having access to blocks had an average milk yield of 20.7 kg/day, which was 1.3 kg (6.7 per cent) higher ($P < 0.01$) than the average of the control group. Additional advantages of blocks included increased conception rate (12.2 per cent), decreased occurrence of diseases (22.5 per cent), improved body condition (Chen *et al.*, 1992) and increased income.

Urea-MNBs was given to Holstein cows in mid lactation by Xu *et al.* (1993). Cows produced 20.5 kg of milk, which was 4.1 kg (25 per cent) higher than the average of the control group. It was estimated that cows consuming the MNBs increased income by ¥ 736 per head per year.

Sheep and Goats

Xu *et al.* (1994) observed increased intake and improved daily gain (26.5 per cent) in sheep having access to MNBs, compared to control animals. They also produced better quality wool with higher S and mineral content. Similar results were observed by Yang *et al.*

When hybrid goats had access to urea-MNBs for two months, average intake of the blocks was 39.5 g/day. Daily weight gain for goats was 85 with and 62 g/day without MNBs. Net income was increased by ¥ 10.78 with blocks. Liu *et al.* (1995) reported results with goats, which grazed on hill pasture during the day and were offered rice straw *ad libitum* in stalls at night. Goats with free access to urea-MNBs along with rice straw at night performed better than those in the control group. Gains were significantly higher in animals with blocks (95 *vs* 73 g/day). The effects of MNBs on performance of growing goats were investigated by Zhang *et al.* Goats with blocks had weight gains 38.3 per cent higher than those without.

Buffaloes

Effects of feeding blocks to buffaloes have been observed by some workers. When buffalo heifers on rice straw diets were supplemented with urea-MNBs, daily gains were 650 g, compared to 620 g for control animals. Feed cost and concentrate consumption per kilogram of gain were 9.82 and 33.3 per cent lower for supplemented buffaloes than those without. Animals showed no signs of poisoning despite block intakes above 1.0 kg/d, indicating that the blocks were safe. Zou *et al*. selected a formula of MNBs for growing buffaloes that contained molasses, urea, grain by-products, minerals and vitamin premix. Intake of the blocks increased with time, and was 172.4, 330.2 and 374.1 g/day at 30, 60 and 80 days, respectively, from the start of the experiment. Compared to control animals, buffaloes with blocks showed higher weight gain (22.6 per cent; 395.4 *vs* 484.6 g/day) and used 22.5 per cent less feed and 22.8 per cent less concentrate per kg of gain.

CEREAL STRAWS AND STOVERS

Cereal straws and stovers may be used as the only feed or as part of the ration, depending on animal type and physiological stage. For the purpose of this technology, ruminants may be classified into two categories: meat animals and dairy cows.

MEAT TYPE (BEEF CATTLE, HEIFERS, SHEEP AND GOATS)

Many studies have been carried out to study the effect of ammoniated straw as the only roughage or as a large proportion of the diet of meat animals. In a trial with yearling steers, Xu compared the effects on animal performance of different levels of untreated and ammoniated wheat straw as the sole roughage. When untreated straw was offered at a high level (60 per cent), daily gain was lowest due to poor total intake. Concentrate consumption per kg of gain was high, even at the maximum straw level, but it was much lower at the highest level of inclusion of ammoniated straw, indicating a saving effect.

Liu *et al.* (1991a) investigated the effect of substituting ammonium-bicarbonate-treated rice straw for hay in growing heifers supplemented with 10 kg maize silage or fresh Chinese milk vetch, plus 2 kg of concentrate. Heifers consumed more DM from ammoniated straw (5.1 kg/day) than from hay (4.0 kg). They had significantly higher daily gains (780 vs 578 g/day; $P < 0.01$). Similar results were obtained with ammoniated wheat straw in young bulls and growing lambs, and with ammoniated rice straw in buffaloes and Holstein heifers, steers, and heifers. It can be concluded that ammoniated straws and stovers can be given to beef cattle, heifers, sheep and goats as sole roughage or as a major part of the diet.

DAIRY COWS

There are few reports on ammoniated straw for dairy cows, since it is widely believed that feeding it to high yielding cows may have adverse effects on milk production. However, the limited reports indicate that dairy cows perform well when they consume ammoniated straw.

Li and Wu compared straw intake and milk production in dairy cows given either untreated or anhydrous-ammonia-treated rice straw. Daily intake was

4.58 and 6.24 kg, and daily milk production 13.6 or 16.0 kg for untreated or ammoniated straw, respectively. In a trial with the ammonium-bicarbonate-treated rice straw, Liu *et al.* found that 17-19 kg of milk could be obtained from Holstein cows when the ammoniated straw represented half of the intake (85 per cent of roughage). These results were consistent with those of Ørskov, where daily milk yield was 23.5-27.4 kg when ammoniated straw, as the sole roughage, was 35-60 per cent of the diet.

It can be concluded that ammoniated straw is a good source of roughage for dairy cows, completely or partially substituting hay. However, it is certainly advisable to use other roughage sources in order to ensure a complete diet.

SUPPLEMENTATION WITH CONCENTRATE AND PROTEIN

The first constraint for low quality crop residue use is usually N. It is desirable that supplementation ensures an almost continuous supply of ammonia-N. Urea is commonly used as a source of fermentable N, and can be sprayed on cereal straw or be mixed with energy supplements. At present, non-protein N supplements have not been widely fed by farmers in China, but urea is usually used as a source of ammonia to upgrade crop residues. With the development of animal production based on crop residues, ammoniation of cereal straw has been widely extended all over China.Since fermentable N is not limiting in ammoniated straw diets, greater consideration should be given to rumen-protected amino acids. They may overcome specific deficiencies limiting production or may be catabolized to improve the supply of glucogenic substrates, which are also usually insufficient in straw-based diets. Protein supplements used in China are mainly oilseed cakes or meals, such as cottonseed cake (CSC) and rapeseed meal (RSM). Sometimes farmers provide their animals with by-products such as rice and wheat bran, or home-grown or self-mixed concentrate mixtures.

Effectiveness of Protein Supplementation

Typical data on animal performance. In Henan Province, an experiment with 40 young Chinese Yellow bulls was conducted to investigate CSC supplementation levels with urea-treated wheat straw. The cattle without CSC gained only 250 g/day, but with CSC gained significantly more ($P < 0.01$). There were no significant differences between 2 and 3 kg, and between 3 and 4 kg of CSC, but there was between 1 and 2 kg. DM intake of wheat straw did not decrease when 1 kg CSC was supplemented, but it did at higher levels. The economic analysis indicated that a relatively high level of production and the highest profit were obtained with 2 kg of CSC.

Similar results were obtained in Hebei Province. Supplementation with 0.25 kg CSC significantly improved feed utilization efficiency and feed efficiency (kg of feed DM/kg gain being reduced from 53 to 15). The group given 1.5-2.0

kg of CSC gained approximately 800 g/d, with only 8 kg feed consumed per kg gain. Further increases of CSC resulted in slightly faster gains at the expense of straw intake.Formaldehyde treatment was effective in improving protein use efficiency of RSM. In terms of nutrient digestibility and N use, formaldehyde-treated RSM compares favourably with treated soybean meal when given to sheep fed ammoniated rice straw. When an ammoniated rice straw-based diet was supplemented with 1.2 kg RSM, heifers had lower weight gains, whereas feeding a similar level of treated RSM resulted in considerably faster gains (15 per cent on average) along with improved feed conversion ratio (FCR) and decreased feed cost.

RESPONSE OF SUPPLEMENTATION WITH UNTREATED AND TREATED STRAW

There are consistent responses in performance to supplementation with protein or concentrate, but the effects are more pronounced when the straw has been chemically treated.Meng and Xiong studied the effect of supplementing with different levels of concentrate on growth rate of Wuzhumuqin wether lambs fed untreated or ammonia-treated wheat straw. Concentrate mixtures were chosen to obtain minimum feed cost per unit gain.Intake of both untreated and ammoniated straw diminished with increasing levels of concentrate, while animals consumed more ammoniated straw than untreated. Ammonia treatment increased straw DM intake on average by 72, 51, 57 and 117 per cent, resulting in higher total DM intake for ammoniated straw diet.

Treatment of wheat straw with ammonia resulted in a significant improvement in weight change of sheep and in feed conversion. Supplementation with concentrate increased weight gain of wethers offered either untreated straw or the ammoniated, but the nature of the response differed according the type of straw. In order to obtain similar daily gains, larger amounts of concentrate were needed for untreated wheat straw than for ammoniated diets.

This indicates that benefits of ammoniation of straw are highest when the supplement level is low. The results show that minimum concentrate consumption per unit gain was obtained when the concentrate accounted for 45 per cent of diet based on ammoniated wheat straw.

Similar results were reported by Liu *et al.*, with RSM as a supplement for Huzhou lambs offered untreated or ammoniated rice straw. The optimum level of RSM with ammoniated rice straw was 100 g/day, whereas 200 g/day had to be supplied to obtain the same liveweight gain with untreated straw.

Bibliography

B.P. Uniyal, J.R. Sharma, U. Choudhery and D.K. Singh: *Flowering Plants of Uttarakhand : A Checklist,* Bishen Singh Mahendra Pal Singh, Delhi, 2007.

Balbas: *Recombinant Gene Expression: Reviews and Protocols*, Springer, Delhi, 1995.

Benjamin Minge Duggar: *Fungous Diseases of Plants*, Agro Botanical Sciences/anica, 1998.

D.P. Dhoundial: *Records : Geological Survey of India : Volume: 123, Part 1: Annual General Report 1988-89*, Geological Survey of India, 2002.

Gyan Deep Singh: *Genetic Engineering of Plants*, Anmol Publication, Delhi, 2008.

Hema Gautam: *Genetic Modification of Plants*, Rajat Publication, Delhi, 2006.

K. Krishnanunni: *Records of the Geological Survey of India : Vol. 133 Part-1 : Annual General Report 1998-1999*, GSI, 2001.

K.R.S. Sambasiva Rao and K. Ananda Krishna: *Regenerative Medicine : Stem Cells and Their Applications*, I.K. International Publishing House, Delhi, 2010.

Kamal Krishna Joshi and Sanu Devi Joshi: *Genetic Heritage of Medicinal and Aromatic Plants of Nepal Himalayas*, Buddha Academic Publication, Delhi, 2001.

M.S. Bista, M.K. Adhikari and K.R. Rajbhandari: *Flowering Plants of Nepal (Phanerogams)*, Department of Plant Resources, National Herbarium and Plant Laboratories, 2001.

N. Sasidharan: *Flowering Plants of Thrissur Forests (Western Ghats, Kerala, India)*, Scientific Publication, Jaipur, 1996.

P. Kachroo: *Progress in Cytogenetics : Prof. A.K. Koul Commemoration Volume*, BSMPS, Delhi, 1999.

P.C. Pande, Lalit Tiwari and H.C. Pande: *Folk Medicine and Aromatic Plants of Uttaranchal*, Bishen Singh Mahendra Pal Singh, Delhi, 2006.

P.M. Tejale: *Records of the Geological Survey of India, Vol. 140, Part-1. Annual General Report 2005-2006*, Geological Survey of India, 2008.

Premmohan M Salwan: *Question Bank With Model Test Papers On General Economics*, Taxmann, Delhi, 2007.

R C Upadhyaya: *Genetics of Flowering Plants*, Anmol Publication, Delhi, 2008.

R. S. Aggarwal: *Quick Learning Objective General English*, S. Chand Publisher, Delhi, 2006.

Rabindra Narain and Surendra Naha: *Genetic Engineering in Plants*, Dominant Publication, Delhi, 2006.

S. Chidambaram, K. Srinivasamoorthy, R. Manivannan, P. Anandhan and Al. Ramanathan: *Recent Trends in Water Research : Remote Sensing and General Perspectives*, I.K. International, Delhi, 2010.

S. Sundara Rajan: *Principles of Molecular Genetics*, Anmol Publication, Delhi, 2003.

S. Thirugnanakumar, K. Saravanan, N. Senthilkumar, A. Anandan and R. Eswaran: *Quantitative Genetics and Crop Breeding*, New India Publishing Agency, Delhi, 2012.

S.C. Agarwal and R.N. Pati: *Folk Medicine, Folk Healers and Medicinal Plants of Chhattisgarh*, Sarup Book Publication, Delhi, 2010.

S.C. Dey: *Flowers from Bulbous Plants*, Agrobios Publication, Delhi, 2003.

Sankar Kumar Bhaumik: *Reforming Indian Agriculture : Towards Employment Generation and Poverty Reduction: Essays in Honour of G.K. Chadha*, Sage Publication, Delhi, 2008.

T.K. Bose, B. Chowdhury, P. M. Cameron, G.G. Maiti and R.G. Maiti: *Garden Plants in Colour : Trees - Tropical and Subtropical*, Naya Udyog, Delhi, 2004.

Verne Grant: *Genetics of Flowering Plants*, Bishen Singh Mahendra Pal Singh, Delhi, 2004.

Zamir K. Punja: *Fungal Disease Resistance in Plants : Biochemistry, Molecular Biology and Genetic Engineering*, I.K. International, Delhi, 2005.

Index

I

K

L

M

N

O

P

S

T

V